工业设计专业系列教材

中英双语工业设计

Industrial Design in both English and Chinese

江建民 毛荫秋 毛溪 编著

中国建筑工业出版社

图书在版编目(CIP)数据

中英双语——工业设计/江建民，毛荫秋，毛溪编著．—北京：中国建筑工业出版社，2009
(工业设计专业系列教材)
ISBN 978-7-112-11333-0

Ⅰ.中… Ⅱ.①江…②毛…③毛… Ⅲ.工业设计-双语教学-终生教育-教材-汉、英 Ⅳ.TB47

中国版本图书馆CIP数据核字(2009)第169187号

责任编辑：李晓陶　李东禧
责任设计：赵明霞
责任校对：兰曼利

工业设计专业系列教材
中英双语——工业设计
Industrial Design
In both English and Chinese
江建民　毛荫秋　毛溪　编著
*
中国建筑工业出版社出版、发行(北京西郊百万庄)
各地新华书店、建筑书店经销
北京天成排版公司制版
北京云浩印刷有限责任公司印刷
*
开本：787×1092毫米　1/16　印张：14¼　字数：356千字
2009年12月第一版　2017年2月第二次印刷
定价：42.00元
ISBN 978-7-112-11333-0
(18546)

工业设计专业系列教材　编委会

序

工业设计学科自20世纪70年代引入中国后，由于国内缺乏使其真正生存的客观土壤，其发展一直比较缓慢，甚至是停滞不前。这在一定程度上决定了我国本就不多的高校所开设的工业设计成为冷中之冷的专业。师资少、学生少、毕业生就业对口难更是造成长时期专业低调的氛围，严重阻碍了专业前进的步伐。这也正是直到今天，工业设计仍然被称为“新兴学科”的缘故。

工业设计具有非常实在的专业性质，较之其他设计门类实用特色更突出，这就意味此专业更要紧密地与实际相联系。而以往，作为主要模仿西方模式的工业设计教学，其实是站在追随者的位置，被前行者挡住了视线，忽视了“目的”，而走向“形式”路线。

无疑，中国加入世界贸易组织，把中国的企业推到国际市场竞争的前沿。这给国内的工业设计发展带来了前所未有的挑战和机遇，使国人越发认识到了工业设计是抢占商机的有力武器，是树立品牌的重要保证。中国急需自己的工业设计，中国急需自己的工业设计人才，中国急需发展自己的工业设计教育的呼声也越响越高！

局面的改观，使得我国工业设计教育事业飞速前进。据不完全统计，全国现已有几百所高校正式设立了工业设计专业。就天津而言，近几年，设有工业设计专业方向的院校已有十余所，其中包括艺术类和工科类，招生规模也在逐年增加，且毕业生就业形势看好。

为了适应时代的信息化、科技化要求，加强院校间的横向交流，进一步全面提升工业设计专业意识并不断调整专业发展动向，我们在2005年推出了《工业设计专业系列教材》一套丛书，受到业内各界人士的关注，也有更多的有志者纷纷加入本系列教材的再版编写的工作中。其中《人机工程学》和《产品结构设计》被评为普通高等教育“十一五”国家级规划教材。

经过几年的市场检验与各院校采用的实际反馈，我们对第二次8册教材的修订和编撰，作了部分调整和完善。针对工业设计专业的实际应用和课程设置，我们新增了《产品设计快速表现诀要》、《中英双语工业设计》、《图解思考》三本教材。《工业设计专业系列教材》的修订在保持第一版优势的基础上，注重突出学科特色，紧密结合学科的发展，体现学科发展的多元性与合理化。

本套教材的修订与新增内容均是由编委会集体推敲而定，编写按照编写者各自特长分别撰写或合写而成。在这里，我们要感谢参与此套教材修订和编写工作的老师、专家的支持和帮助，感谢中国建筑工业出版社对本套教材出版的支持。希望书中的观点和内容能够引起后续的讨论和发展，并能给学习和热爱工业设计专业的人士一些帮助和提示。

2009年8月于天津

前　言

虽然十年前就与何人可、张兵一起合作参加过“普通高等工科教育机电类规划教材”《工业设计专业英语》的编写，但总觉得现今的专业英语教材脱不开普通英语的套路，教材的面孔很强，往往较少考虑专业本身的系统及专业语言表达上应有的特点。在这种潜意识的支配下，到手头积累了自认为足够多的材料时终于就动手干开了，把过去编写专业英语教材的甜酸苦辣忘了个一干二净。

本书每课课文的正文按照工业设计专业的知识结构需要与实际设计流程分成了产品造型、概念设计、系统设计、产品企划、产品开发以及新设计理念6个单元。这6个单元、24课可以说包容了当今工业设计领域的众多重大话题。每个单元包含的4篇课文，选择了该单元最重要的话题。

每课除了按常规列有单元序号和课文序号、课文标题、词汇和短语，以及课文正文(中英文对照，并包括了与课文正文配套的必要图片资料)外，每课课文正文前面还有一般教材上少见的“词汇联想与记忆”以及“关键词汇和概念”两个栏目，以期在这两方面对读者有所帮助。

为了弥补每课课文正文容量的不足，每课课文之后还增加了“拓展阅读”栏目。故在书尾除附有全书的词汇表外，还附有全书各课拓展阅读的参考中译文。

为了调节“气氛”，免得全书内容过于偏重理论而使读者感到沉重，在课文正文与拓展阅读中，选用、介绍和展示了大量现今世界上最重要的设计公司、他们的著名设计师及其典型设计作品。所有文字都尽量配上相应的图片，以期使本书成为理论性强、趣味性浓、英语纯正、图文并茂，工业设计专业和英语两方面俱佳的入门书籍。

江建民

2009年7月于无锡

目　录

第1单元
Unit 1

第1课 | 魅力和产品造型(上)
Lesson 1 | Attractiveness and product style(Part I)

词汇和短语
Words and phrases

attractiveness　魅力

grab　抢夺，夺取，攫取

grab one's attention　引起注意

flick　轻弹，突然移动；(瞬间)照见

flick over　翻阅

product marketing　产品营销

visual appearance　视觉外观

literal meaning　字面含义

prior knowledge attractiveness　前理解的魅力

more than half the battle　成功了一多半

suffer a frustration　遭受挫折

make sense　有意义

repeat sales　重复销售

predecessor　前任，前辈

fail to do so　未能这么做

词汇联想与记忆
Association and memory of words

attract *vt.*　吸引

vi. 有吸引力，引起注意

market *n.* 市场，销路，行情

vt. 在市场上交易，使上市

vi. 在市场上买卖

vision *n.* 视觉；先见之明，眼力；想象力，幻想，景象

vt. 梦见，想象，显示

sense *n.* 官能，感觉，判断力，见识；意义，理性

vt. 感到，理解，认识

关键词汇和概念
Key words and concepts

魅力，就是吸引力，使人渴望拥有动人、诱人和愉悦的能力。

The attractiveness is having the power attractive，and the power to make desirable and pleasing to somebody.

产品造型是指一个产品总的外表、模样，或设计。

Product style means general appearances，form，or design of a product.

产品造型设计是赋予产品魅力的主要手段之一。

Product styling is one of key methods to give a product attractiveness.

课文
Text

魅力和产品造型(上)
Attractiveness and product style(Part Ⅰ)

魅力这个词极好地描述了消费者眼中产品应有的特征，因而原谅了它用词本身的不雅。产品应该以三种有极细微差异的方式对消费者显现出吸引力:

Attractiveness is itself an inelegant word. But it describes so perfectly what products should be in the eyes of a customer that its inelegance is forgiven. Products should be attractive to customers in three subtly different ways:

● 第一， 一个物体如果能说引起了你的注意(当然是因视觉上的愉悦而不是非常丑陋)，就可以说它是有魅力的。就像建筑师谈到房子时说它有“街沿石”的吸引力一样。在你步行或驾车经过一栋设计得很好的房子时，它的外观形象应该从街沿石起就开始吸引人。当你在商店里

从它们边上徒步经过或者翻阅商业传单看到它们的照片时，它们应该能立即引起你的注意。

● Firstly，an object can be said to be attractive if it grabs your attention(by being visually pleasing，of course，rather than outstandingly hideous). Architects refer to houses as having "kerb appeal". When you walk or drive past a well-designed house，its visual appearance，from the kerb，should be immediately attractive. They should immediately grab your attention as you walk past them in a shop or as you flick over their photograph in a leaflet.

● 第二，有魅力的物体应令人渴望拥有。从产品营销上讲，让消费者想要拥有某个产品就是成功了一大半。如果一个产品能够做到仅由其视觉外观就能引起消费者的愉悦，这实际上就是非常有力的产品营销。

● Secondly，an attractive object is a desirable object. Making customers want to own a product is more than half the battle in product marketing. If a product can be made desirable to customers simply by its visual appearance，then this is powerful product marketing，indeed.

● 将这两种含义归并到一起，即一个产品既能引起关注又能使人愉悦，就意味着消费者被吸引得“挨近”这个产品了，这就是“魅力”这个词字面上的含义。

● Taking these two meanings together-a product which is both attention grabbing and desirable-means that customers are "drawn towards" the product; the literal meaning of the word attractive.

如果一个产品由其外观引起消费者欲求，实际上这就是有力的产品营销
If a product can be made desirable to customers simply by its visual appearance，then this is powerful product marketing，indeed

魅力的四种情况(上)

The four faces of attractiveness(Part I)

为了使产品具有魅力，我们需要懂得产品能被消费者认为有吸引力的方法。这主要有四种途径。

In order to make products attractive，we need to understand the way in which products can be seen as attractive by customers. There are four main ways.

1．前理解的魅力

1．Prior knowledge attractiveness

大多数人像我一样都经历过这样的挫折：买了一本小说，不料想发现几年前就已经买过和读过，它只不过是用了全新的封面设计。在出版界这是有商业意义的，但在产品设计上这却是个灾难。很多类型的产品，其商业上的成功取决于重复销售。如果这样一个产品在再设计时产品的外观发生了根本的改变，以至于消费者无法辨认出它就是以前用过而且喜欢的产品，那么寻找重复销售的前景就被毁掉了。所以，如果你正在设计的产品是对现有产品的改进，在视觉识别方面保持其外观与其“前辈”的同一性就很重要。否则就可能危及到现有消费者的重复销售。如果这个产品是一个系列中的一个，或者是以同一个品牌或同一个公司形象下进行的销售，就必须让潜在的消费者能从其外观形象上很明显地看到这一点。即使这是个独特而又全新的产品，消费者也必须能从其外观形象上识别出这是哪类产品。例如，要是消费者在商店里在一台草坪割草机边上走过时都没能认出它是割草机，那么即使这是台最具创新性和极富想象力的新型草坪割草机，生产它也就没有任何意义。这一点对产品造型很重要，但又常常被设计师所忽视，或者认为对造型创新施加了不适当的限制而予以抱怨。从营销和商业的观点看，这个问题是生死攸关的，在造型过程中必须得到认可和重视。

The most people, like me, have suffered the frustration of buying a novel only to find that it is one you had bought and read years ago when it had a completely different from cover design. This makes business sense in the publishing business but it can be a disaster in product design. Many types of products are dependent upon repeat sales for their commercial success. If, during a redesign of such a product, its visual appearance is so radically changed that customers no longer recognize it as the product they previously used and liked then the prospects for repeat business are destroyed. So, if the product that you are designing is an update of an existing product it is important to maintain the visual identity of its predecessor. To fail to do so could jeopardise prospects for repeat purchases by existing customers. If the product is a part of a product range or is sold under a brand or company identity, that must be made obvious to potential customers from its visual appearance. Even if it is unique and completely new product, customers must be able to recognise what kind of product it is from its visual appearance. There is no point in producing the most innovative and imaginatively designed new lawn mower, for example, if customers are going to walk past it in the shop, not realizing it is a lawn mower. This is an important aspect of product styling which is often overlooked by designers or resented as an undue constraint on styling innovation. From a marketing and business point of view, it is vitally important and it must be acknowledged and respected during the styling process.

很多产品的商业成功依赖于它们的重复销售
Many types of products are depeudent upon repeat sales for their commercial success

拓展阅读
Extend and reading

Nokia 2650 mobile phone

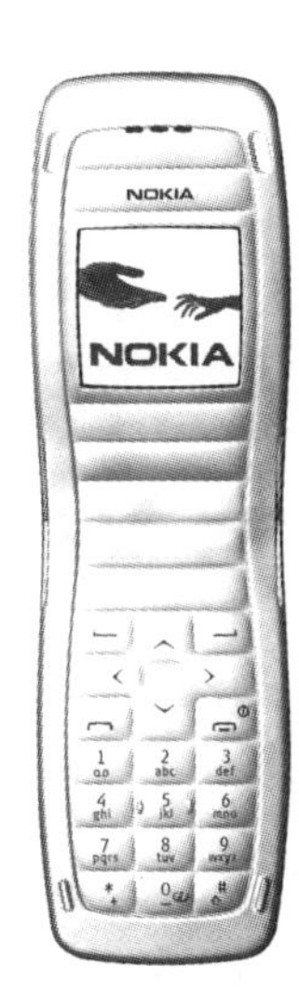

The key innovation of the Nokia 2650 mobile phone is its unique folding mechanism. The phone doesn't use a hinge, like competitors' models, resulting in a uniquely unified look in both the open and closed positions. Hard on the outside and soft on the inside, the 2650 is tough and durable when closed but opens to a soft, pillowed surface for the user's face.

The design of the 2650 embraces the qualities of balance, symmetry and simplicity, making it less of a high-tech, complicated device and more of a straightforward, useful product. In additional to its unique folding mechanism, the 2650 is also much simpler and less costly to manufacture than other folding phones.

Nike considered boot

The Nike considered boot is a performance shoe that combines subtle styling with unique environmental benefits without sacrificing Nike's commitment to design innovation. The handcrafted boot sports a unique look obtained by weaving hemp lace between the leather upper to produce a one-of-a-kind shoe that molds to the foot. The project grew out of consumer feedback that clearly indicated a desire for more sustainable products. The designers achieved impressive environmental statistics: a reduction of 61 percent in manufacturing waste, 35 percent in energy consumption and 89 percent in the use of solvents. In addition, the shoe is designed to be more effectively recycled in Nike's Reuse-A-Shoe program. The product launch has surpassed expectations and has led Nike to consider similar issues in its other product lines.

"An amazing confluence of handcraft and high technology."

"This boot was a big winner because it hit all the design chords: looks good, looks human, looks crafted, looks ecological, looks functional-the leap, looks fashionable! Nike transformed granola and broccoli into voguish shoes!"

iXi bicycle

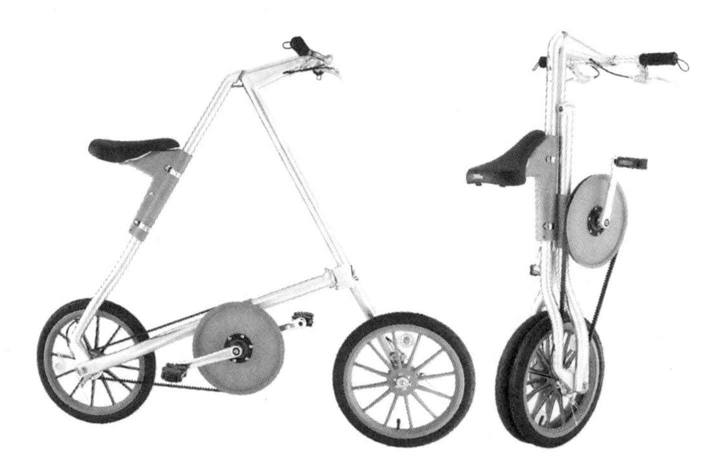

The goal was to design a bicycle that would be visually appealing and user friendly in the manner of other modern consumer durables. It should appeal to sophisticated consumers who might not normally venture into a traditional bike store. At the same time, the bike should function in a way that people can easily integrate into their daily lives, be extremely easy to use, not only for riding, but when not in motion. The Drive Belt system is entirely maintenance free, requiring no tensioning or lubrication. Media attention following the launch of the bike dramatically raised the company's profile resulting in easier sales and renewed attention for the entire product line within industry.

“RKS Pop” series guitars

The Pop series guitar is a high quality, innovative and versatile instrument that is available in a variety of eye-catching colors. The design strategy focused on two main issues: the lack of recent innovation in the guitar industry and the consumption of irreplaceable natural resources in the production of guitars. The Pop series guitars have been perfected for balance, ergonomics and tone. A wide assortment of ribs and bodies in many bold, contemporary colors can then be added to produce an infinite variety of styles to appeal to a wide array of consumers. From a sustainability standpoint, the amount of tone woods has been significantly reduced. Typically guitar contains eight to ten pounds of tone woods from the rain forests of Central and South America. The RKS Pop series uses less than two pounds woods which is obtained from domestic tree farms.

第2课 | 魅力和产品造型(下)
Lesson 2 | Attractiveness and product style(Part Ⅱ)

词汇和短语
Words and phrases

functional attractiveness　功能魅力

functional value　功能价值

product semantics　产品语义学

product meanings　产品的含义

symbolic attractiveness　符号魅力

product' s symbolism　产品符号学

party animal　社交(动物)一族

village community　乡村社区

theme board　主题看板

inherent attractiveness　固有的魅力

aesthetic appeal　美学诉求

product planning　产品企划

concept design　概念设计

词汇联想与记忆
Association and memory of words

perception　*n.*　知觉，知觉力

sign　*n.*　标记，符号；记号，迹象，征候

　　v.　签名(于)，署名(于)；签署

style　　*n.*　风格，时尚，文体，风度，类型
　　　　vt.　称呼，设计，使合潮流
appeal　*n.*　请求，呼吁，吸引力；要求
　　　　vi.　求助，诉请，要求

关键词汇和概念

Key words and concepts

产品的魅力有四种：前理解的魅力、功能魅力、符号魅力，和视觉形式固有的魅力。它们各有自己的内涵和背景。

There are four faces of attractiveness for products: prior knowledge attractiveness, functional attractiveness, symbolic attractiveness and inherent attractiveness of visual form. They have their own meanings and backgrounds.

课文

Text

魅力的四种情况(下)

The four faces of attractiveness(Part II)

2．功能魅力

2．Functional attractiveness

要是消费者对于产品缺乏前理解，产品的外观形象就必须用某种方式激发和启示他们在以往的使用经历中提供的信心。对于功能价值很重要的这类产品，可以用这种办法来实现：让产品看起来会很好地执行其功能。产品看起来会很好地执行功能跟产品实际上工作得很好是不一样的。消费者在购买之前常常没有机会充分试用一个产品，因此他们对产品性能的许多判断是基于该产品看起来将如何运行。整个产品设计部门都要围绕着这样一个原则来进行开发：使产品看起来能够很好地履行其设定的用途。这称为产品语义学：从字面上讲就是产品的含义。

For customers with no prior knowledge of a product, visual appearance must somehow inspire the confidence that prior use would otherwise provide. For products in which functional value is of importance, this is achieved by making the product look like it will perform its function well. This is different from actually making the product work well. Very often customers will not have the opportunity to thoroughly test a product before buying it. A great deal of their judgement on performance is, therefore, based on how the product looks like it will function. A whole branch of product design has developed around the principle of making products look like they perform their intended purpose well. It is known as product semantics: literally product meanings.

很多顾客对产品表现的判断来源于它们看上去的功能
A great deal of the cusfomers judgement on performance is based on how the product looks like it will function

3. 符号魅力

3. Symbolic attractiveness

当外观价值成为购买该产品的重要(甚或是全部)理由时，就要求其产品造型用不同的方法。在此，产品的符号学就很重要。购买信心的激发来自于产品能反映消费者自身形象的程度以及他们希望产品在别人眼中反映出的看法。设计师能够采用使产品具有某种符号表达的办法而使产品的外观具体化。例如可以包括："这个产品非常有趣，而我就是这社交一族"，"这是个典雅而传统的产品，而我就是这乡村社区的栋梁"，或者"这是个很高尚而有点儿大胆的产品，而我还不像我女儿认为的那样老"。在设计过程中，消费者企求的这些陈述和想象可以通过采用生活方式、情感和主题看板等方法予以发展和贯穿起来。

Where appearance-value is an important part(or all)of the reason for purchasing a product, a different approach to its styling is required. Here the product's symbolism is important. Purchasing confidence is inspired by the extent to which the product reflects the customer's self image and the statement that they wish the product to make in the eyes of others. Designers can embody these in the appearance of the product by having it make some sort of symbolic statement. Examples might include, "This is a fun product and I am a party animal", "This is a refined, traditional product and I am a pillar of the village community", or "This is a very respectable but slightly risky product and I am not as old as my daughter thinks I am". These statements and the images they conjure are developed and articulated during the design process by the use of lifestyle, mood and theme board.

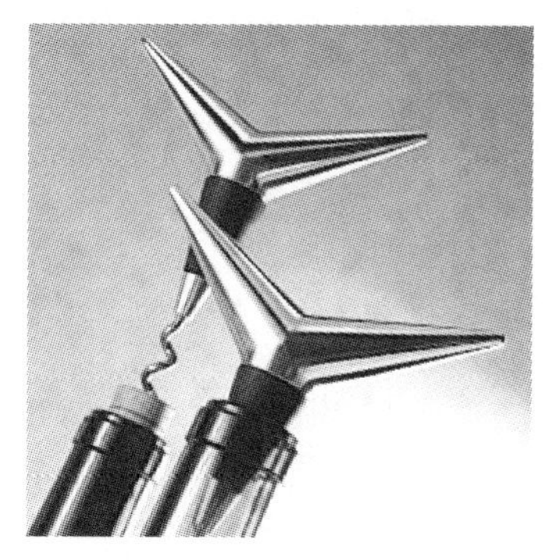

产品在消费者自身以及在别人眼中的形象反映程度会激发购买的信心
Purchasing confidence is inspired by the extent to which the product reflects the customer's self image and the statement they wish the product to make in the eyes of others

4. 视觉形式固有的魅力

4. Inherent attractiveness of visual form

对于任何一类产品，外观形象根本上是极难明了、难以捉摸的品质：它的高雅，它的美丽，它的内在的美学诉求。正如上面叙述的，这是产品魅力在感知、社会和文化层面上的决定性的体现。

At the root, visual appearance, for a product of any sort, is the most elusive and intangible quality: its elegance, its beauty and its intrinsic aesthetic appeal. This is the embodiment of the perceptual, social and cultural determinants of the attractiveness of products, as described above.

造型不是在任何一个单点上的及时注入，也不是在最后加上去的某种东西。造型贯穿于整个设计过程。

Styling is neither injected at any single point in time, nor is it something added on at the end. Styling continues throughout the design process.

这些魅力因素最终就是产品本质性的魅力——由它们在"眼睛诉求"的基本感知来决定的魅力。问题仍然是我们是否能系统地、有序地设计产品使之具有视觉魅力。

The last of these attractiveness factors is the intrinsic attractiveness of products—the attractiveness which is determined by their basic perceptual "eye-appeal". The question remains as to whether we can systematically and methodically style products to be visually attractive.

- 产品企划中就要关注如何能够研究和确定造型的目标。
- Product planning will look at how styling objectives can be researched and specified.
- 概念设计中要检查这些造型目标如何能用情感和主题看板转译为视觉主题，并进而转化为新产品的首个造型概念。

● Concept design will examine how these styling objectives can be interpreted into visual themes by the use of mood and theme boards and then translated into the first styling concepts for the new product.

● 然后显现的是，这些视觉概念怎样转化为能进行建模、市场测试及最终作为成品制造出来的具体的设计。

● It will then show how to convert these visual concepts into a physical embodiment which can be modeled, market tested and ultimately manufactured as the finished product.

拓展阅读

Extend and reading

Intuos3：Gold，computer equipment

The Intuos3 pen/tablet system is a state-of-the-art family of integrated，plug-and-play cordless and battery-free computer graphics tools for professional use. Wacom wanted an innovative，approachable and ergonomic design that would visually convey ease of use and regain market share lost to competitors. Extensive user observation and ergonomic research identified many opportunities for product innovation，such as workspace flexibility，intuitive functionality，a streamlined aesthetic and realistic drawing and painting simulation.

The new design gives users the control，precision and flexibility they need to capture their creativity and its ergonomics allow them to draw or paint for extended periods of time without experiencing discomfort. Since its launch in 2004，the Intuos3 has surpassed initial sales expectations，outselling its nearest competitor two-to-one and helping to drive the company profits up almost 11 percent.

Sony Qualia 016 digital camera

The Sony Qualia 016 is an ultra-compact digital still camera that disappears into your palm. Because of its size，the camera is easily carried with you at all time，ready to take a photo in an instant. The Qualia 016 is not just a camera but a compete kit of accessories to optimize your photo-taking experience，such as a viewfinder，wide conversion lens，teleconversion lens，flash unit，video-

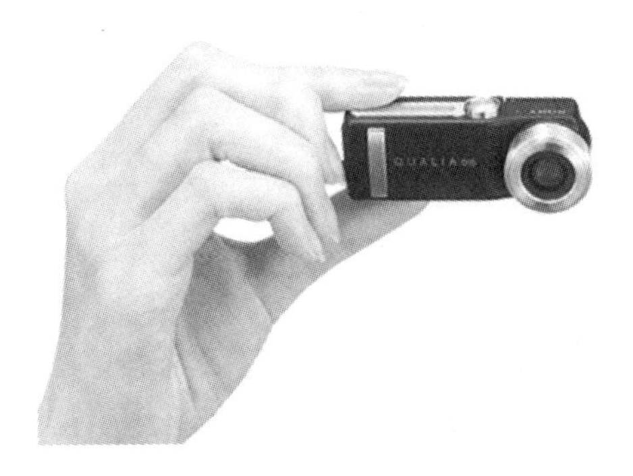

out unit and timer remote unit, which all fit in a custom all-in-one carrying case. However, there is more to this design innovation than its small size. Designers made the camera's interface comfortable and intuitive, over time becoming an extension of your hand and of the company's focus on creating products that appeal to the customers.

"What every well-equipped(and well-heeled)hobbyist-spy wants for Christmas. This cxquisitely detailed ultra-miniature digital camera, about the size of two of your fingers, draws its aesthetic from its finely machined case, lens barrel and a cluster of accessories and attachments which come packaged in a courier-style briefcase. Camera operations are selected by sliding one of those fingers along a linear touchpad."

Spring Roll—Fetch toy for dogs

These injection molded natural rubber dog toys were designed to be enjoyed by both pets and humans. By varying the wall thicknesses, the toys produce an erratic bounce and roll and treats can be placed in a stuffable, hollow core. The toy's thick inner core allows the dog to carry it in a more natural position. The design process also focused on captivating the attention of the customers at the point of purchase: the packing was minimized to a simple wrap that highlights the design. These toys were WETNo2 International's first foray into the dog toy market. The response to date has resulted in plans for an additional line of toys to launch in late 2005.

"Who knew that a fetch toy for dogs could be a colorful piece of sculpture? Given the basic parameters of bounceability, rollability, durability and canine gripability, the designers asked 'why not make it shapeful and beautiful as well?' The result is a delightful exception to the ubiquitous(and rather ugly)rawhide bone."

第3课 | 产品造型的视觉感知(上)
Lesson 3 | Visual perception of product styling (Part Ⅰ)

词汇和短语
Words and phrases

visual perception　视觉感知
visual image　视觉形象
visual signal　视觉信号
visual information　视觉信息
visual styling　视觉形态
visual process　视觉过程
pre-attentive　前期注意
attentive mode　注意模式
pre-conception　前期概念
immediate appeal　直接诉求

词汇联想与记忆
Association and memory of words

rule	*n.*	规则，惯例，章程，标准，控制
	vt.	规定，统治，支配，裁决
symmetry	*n.*	对称，匀称
harmony	*n.*	协调，融洽
form	*n.*	形状，形态，外形，表格，形式
	v.	形成，构成，排列，组成

关键词汇和概念

Key words and concepts

视觉感知是指我们通过视觉开始意识到变化的一个过程。产品造型通常是其视觉形态的一个“缩写”。我们实际看到的只是我们认为我们看到的东西而已。

Visual perception means a process by which we become aware of changes through the sense of sight. Product style is usually an abbreviation for its visual style. We actually see what we think we see.

课文

Text

产品造型的视觉感知(上)

Visual perception of product styling(Part I)

当我们说某一产品很吸引人时，很少指该产品的声音、感觉或味道。这是个明显的暗示，说明人类的感知是由视觉主宰的，而产品造型则通常是视觉形态的一个“缩写”而已。

When we talk of a product being attractive we rarely refer to its sound, feel or smell. This is a striking reminder that human perception is dominated by vision and that product style is usually an abbreviation for visual style.

很明显，大脑对于收到的琐碎的视觉信号进行了巧妙的处理。其最明显的证据就是我们“看到”的视觉印象总是十分连贯和完整的。

Clearly, the brain does some clever processing of the fragmented visual signals it receives. The most obvious evidence of this is that the visual images we “see” are perfectly coherent and integrated.

视觉过程分两个阶段。我们对视觉信息的分析也以两种不连续的方式进行。首先，浏览总体形象，找到模式和形状。这是一瞬间的过程，不要求观察者做到刻意的努力，故可称为“前期注意”。第二步是视觉过程的“注意模式”，包含刻意地集中于印象的细部，审视其组成部分。

Visual process is happened in two stages. Our analysis of visual information takes place in two discrete ways, too. Firstly, the overall image is scanned to look for pattem and shape. This is a rapid process, requiring no deliberate effort on the part of the viewer and is therefore described as preattentive. Secondly, attentive mode of visual processing involves deliberate focusing on details of the image to examine its component parts.

人类的感知是由视觉主宰的，而产品造型则通常是视觉形态的一个“缩写”而已
Human perception is dominated by vision and that product style is usually an abbreviation for visual style

我们总认为，我们的眼睛是世界的窗户。其实不然。我们实际看到的只是我们认为我们看到的东西而已。我们看到一个图像，不加思考，就抽取它的主要特征。从这些特征出发，我们的大脑就形成一个“草图”来说明特征所包括的内容。由这个前期概念引导，注意过程对形象的各组成部分进行研究，并得出细节。

We believe that our eyes are a window on the world. But this is not so. We actually see what we think we see. We look at an image and without thinking, we extract its main features. From these features, our brain works out what it contains by forming a “sketch”. Then attentive processing, guided by this pre-conception, studies the component parts of the image and extracts the detail.

那么，这又告诉我们产品造型是什么呢？我们对造型描述的方式使我们认为，对造型的判断大多是由前注意期的整体处理得到的。我们说产品具有直接诉求，它们有冲击性，能吸引眼球或攫取注意。这些判断是瞬时的，前期注意的；它们并不要求认真刻意的考虑，似乎也不以对产品各组成部分的专注的处理为基础。当我们讲到一个产品的总体形态或形象时，指的就是我们对它的总体感知。因此造型，至少部分地是以前注意期的总体处理为基础作判断的。因而，一个产品的美与我们视觉系统特性的关系，要比跟有关产品固有的美的任何其他因素的关系更为紧密。

So, what does this tell us about product styling? The way we talk about styling suggests that much of our judgment of styling is determined by pre-attentive global processing. We say that products have immediate appeal, and they are striking, eye-catching or attention grabbing. These judgments are instant and pre-attentive. The require no careful deliberation and do not seem to be based on the attentive processing of the product’s component parts. When we talk about the overall form or image of a product, we are referring to our global perception of it. Styling is, therefore, at least partly judged on the basis of pre-attentive global processing. The beauty of a product is, consequently, more to do with the properties of our visual system than anything fundamentally beautiful about the product.

因此造型是，至少部分的是以前注意期的总体处理为基础进行判断的
Styling is, therefore, at least partly judged on the basis of pre-attentive global processing

美，确实是在观看者的眼(及大脑)中。当我们设计一个美的物体时，我们必须将它设计得与人类的视觉感知性能相适应。因此懂得视觉就成了创造美的关键；同时，前期注意视觉感知的规则应该转变成产品造型的原则。

Beauty truly is in the eye(and brain)of the beholder. When we design an object to be beautiful, we must design it to correspond with the perceptual properties of human vision. Understanding vision, therefore, becomes the key to creating beauty and the rules of pre-attentive visual perception should translate into principles of product styling.

我们可以在两个层面上来审视由此得到的视觉感知的规则。首先，是视觉感知的一般规则，它允许我们从任何场景中抽取视觉信息。第二，有一些特定的规则，使我们为了特定的生存原因而对某些视觉任务更加出色地对待。

We can examine these resulting rules of visual perception at two levels. Firstly, there are general rules of visual perception that allow us to extract visual information from any scene. Secondly, there are specific rules which make us particularly good at certain visual tasks, for specific survival reasons.

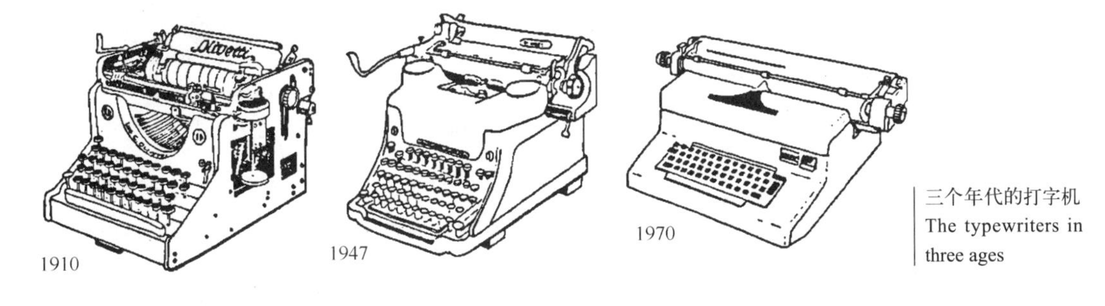

三个年代的打字机
The typewriters in three ages

拓展阅读

Extend and reading

Deluxe gadgets

1. Philip Stein Teslar watch

Ever worry all the technologies you use might be harming your body? Neither do we. But if you are worried about invisible "mind waves", you'll love this watch. It contains a Teslar chip(named after an inventor who worked with Thomas Edison), which emits a signal that screens out electronic pollution, so you can enjoy technology worry-free. Puff Diddy has one, but we're not sure if that's a good thing.

2. Paul Smith shoe care

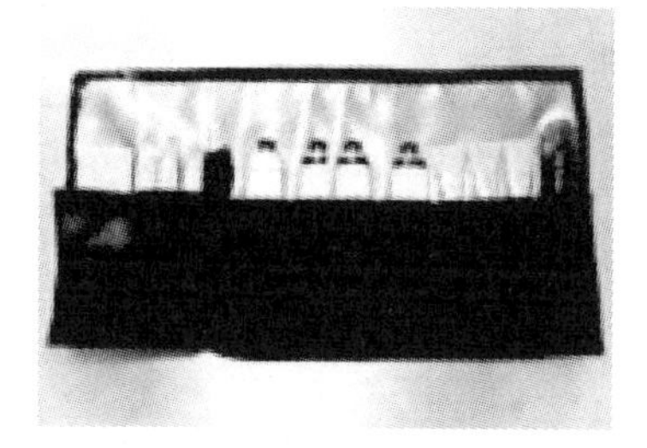

After your face, your shoes are the first thing people notice about you. So even if you're a reasonably handsome chap, your scuffed footwear could let you down. You can avoid embarrassing social incidents with this shoe care kit, which includes two polishing brushes, a cloth, three tubes of different coloured polish, a shoehorn and an application brush. And it all comes in a pouch designed to look like an artists paint set.

3. De Sede chair DS-151

We think you'd better sit down. No, we've not got any distressing news, we just want you to try this amazingly comfortable chair! It's got an adjustable headrest, and the non-gliding stand won't slip when you fall into it at the end of a hard day. Best of all, it's ergonomically sculpted to fit someone lying down with their hands behind their head, doing absolutely nothing. So it suits us perfectly.

4. ZARA 3 Sae baby transporter

The ZARA 3 rolls all of a child's transportation needs for their first three years into one product-from an infant car seat/carrier to traditional stroller and finally a toddler car seat. Its all-in-one composition saves parents'money and its minimal complexity makes daily use and transition between stages easier. Designers eliminated all superfluous elements to increase functionality. The compact chassis makes lifting and stowing in the trunk easier. Textiles became the means for the client to differentiate the ZARA 3 by giving users broader aesthetic choices with ability to change the fabric elements at will throughout the product's life and substituting some mechanical parts with textiles. For instance, the stroller can adopt two positions through regulation of the back and foot stand, which is achieved through textiles by way of zips, Velcro and clips.

第4课 | 产品造型的视觉感知(下)

Lesson 4 | Visual perception of product styling(Part II)

词汇和短语

Words and phrases

gestalt rules　格式塔规则

be predisposed to　倾向于

rule of symmetry　对称性规则

proximity　近似性

similarity　相似性

continuation　连续性

tend to　倾向于，趋于

visual harmony　视觉协调

visual simplicity　视觉简单性

comply with　遵循，遵照

geometric form　几何形状

facial value　外观价值

visual theme　视觉主题

词汇联想与记忆

Association and memory of words

symmetry　*n.*　对称，匀称

simplicity　*n.*　简单，简易，朴素，直率

form　*n.*　形状，形态，外形，表格，形式

v. 形成，构成，排列，组成

关键词汇和概念

Key words and concepts

格式塔心理学的规则告诉我们，我们对对称性和规则形状的察觉有非凡的能力。在察觉某些类型的视觉形式方面，我们的感知能力特别精准。这些，对于产品的造型设计都非常重要。

The gestalt rules tell us that we have a remarkable ability to detect symmetry and regular patterns. Our perceptual ability is particularly well refined in the detection of certain types of visual forms. These are important for product styling.

课文

Text

产品造型的视觉感知(下)

Visual perception of product styling(Part II)

一般规则

General rules

格式塔规则：是20世纪20年代、30年代和40年代一群德国心理学者命名的视觉感知原则。在德语中格式塔的原意即为形式或形状，而格式塔心理学者认为人类的视觉会倾向于看到某些类型的形状。

The "gestalt" rules: it was named by a group of German psychologists working in the 1920's, 30's and 40's. Gestalt is the German word for pattern and the Gestalt psychologists suggested that human vision is somehow predisposed to see certain types of patterns.

最强烈的一条格式塔规则可能要算对称性规则了。我们对于对称性的察觉有非凡的能力，即使是复杂的形状，不完全对称的自然形状，甚至于对称性被严重扭曲了的物体，仍能比较容易地发现它们是否对称。

Probably the strongest gestalt rule is the rule of symmetry. We have a remarkable ability to detect symmetry, though in complex forms, natural forms with incomplete symmetry and even in objects which have had their symmetry substantially distorted, We are still easy to find whether or not they are symmetrical.

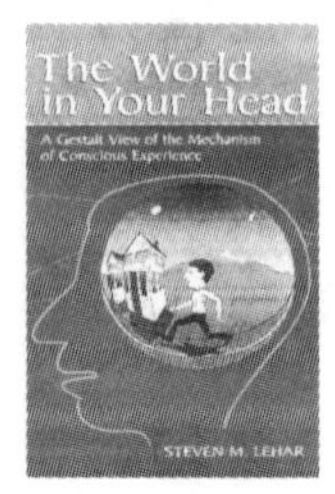

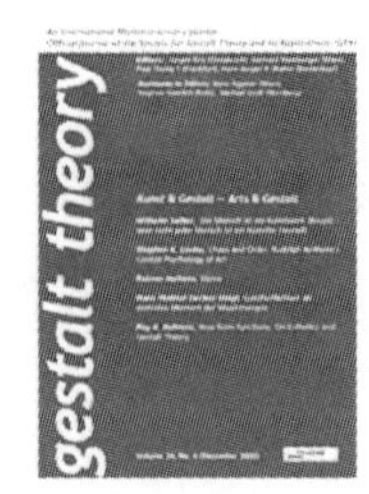

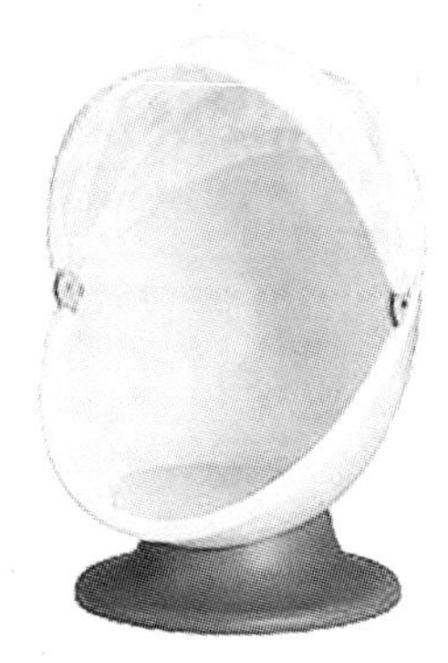

最强烈的一条格式塔规则可能要算对称性规则了
Probably the strongest gestalt rule is the rule of symmetry

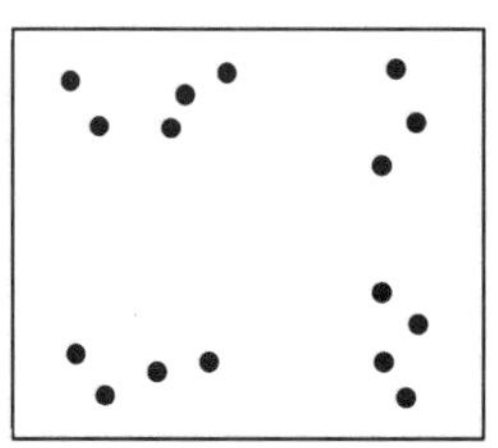

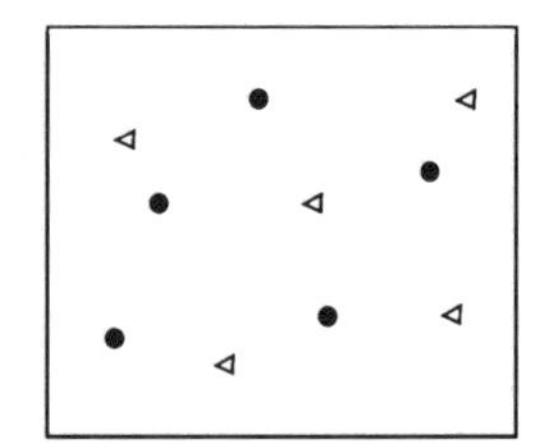

连续性规则
The rule of continuation

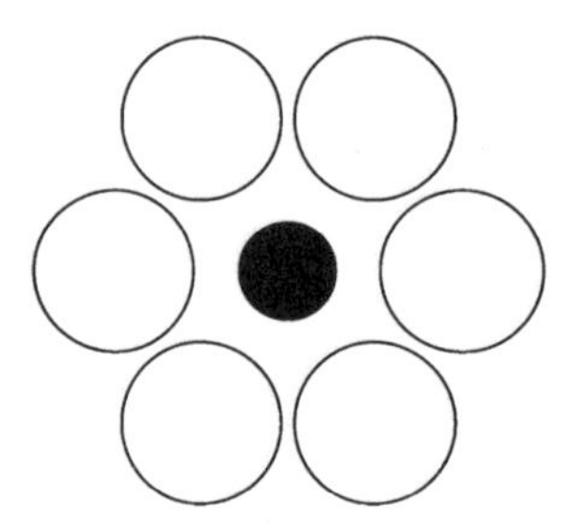

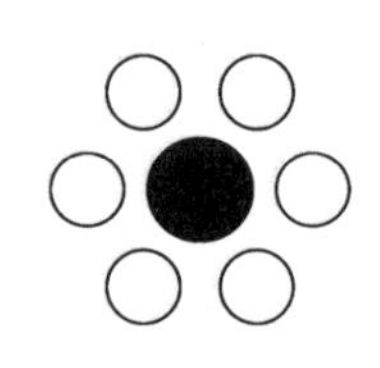

接近性规则
The rule of proximity

相似性规则
The rule of similarity

在感知规则形状方面我们也有杰出的能力，格式塔心理学者将其细分为三个规则：即接近性规则、相似性规则和良好连续性规则。接近性规则提出，物体或特征非常接近的将倾向于看成是同一样式。相似性规则提出，具有相似形状或形式的物体或特征将趋于认为是同一样式。而良好连续性规则提出，形状是因其组成部分的连续性、轨迹或矢量而被感知到的。

We also have remarkable ability to detect regular patterns and the gestalt psychologists break this down to three rules：the rule of proximity，the rule of similarity and the rule of good continuation. The rule of proximity proposes that objects or features which are in close proximity will tend to be seen as a pattern. The rule of similarity proposes that objects or features which are of similar shape or form will tend to be seen as a pattern. The rule of good continuation proposes that patterns are perceived due to the continuity，trajectory or vector of their component parts.

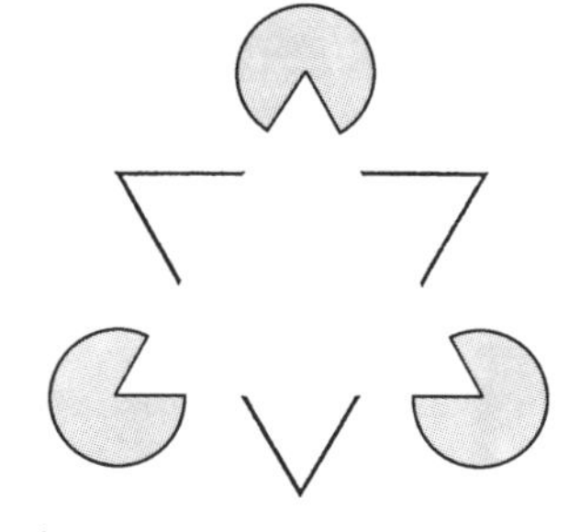

不存在的三角形
A nonexistent triangle

这些规则对产品造型意味着什么呢？

What do these rules mean for product styling?

格式塔规则对产品造型具有深远的意义，而且涉及范围既精深又宽泛。产品各成分或产品特征的有效整合，可以由形状的格式塔规则衍生。功能上有联系的产品特征，可以用这些规则加以组合。

The implications of the gestalt rules for product styling are profound and range from the specific to the sweeping. The effective integration of product components or product features can be derived from the gestalt rules of patterning. Product features which are related functionally，can be made to appear grouped together using these rules.

视觉形式的和谐观念也与格式塔原则有关联。严格地说，格式塔心理学者从未以规则的形式明确阐述过和谐这个观念。但是将视觉简单性的有关规则与视觉形状的有关规则结合在一起就可以认为，视觉的融洽与协调肯定是很重要的。

The notion of harmony in visual forms has also been linked to gestalt rules. Strictly speaking, harmony was never formulated as a rule by the gestalt psychologists. But the rules relating to visual simplicity combined with the rules regarding visual patterns suggest that visual harmony must be important.

视觉简单性

Visual simplicity

格式塔理论对产品造型最透彻的含义就涉及产品的视觉简单性。为遵守强有力的格式塔原则，产品应该是对称的，并由清晰的线条组合起来，构成简单的几何形状。

The most sweeping implication of gestalt theory for product styling concerns the visual simplicity of products. To comply with the most powerful gestalt rules, products should be symmetrical and be comprised by clean lines which go together to make up simple, geometric forms.

视觉过程的特定规则

Specific rules of visual process

首先，在察觉某些类型的视觉形式方面(例如面孔)，我们的感知能力是特别精准的。这使我们在进行产品造型时能开拓这些能力，并根据不同的产品赋予其特定的外观价值。

The first, is that our perceptual ability are particularly well refined in the detection of certain types of visual forms(such as faces). This allows us to exploit these abilities during product styling and bestow certain facial values upon products.

第二，在使用任何视觉主题时，必须十分注意这些主题在人们的价值观中根深蒂固的含义。

The second, enough care must be taken when using any visual theme which has deep rooted meaning in human values.

必须对此类符号学意义进行充分的探究，并在使用它之前必须理解其在人类心理学上的基础。

Any such symbolism must be thoroughly explored and its basis in human psychology must be understood before it is used.

拓展阅读

Extend and reading

Art, design and gestalt theory

by Roy R. Behrens

Abstract: Gestalt psychology was founded in 1910 by three German psychologists, Max Wertheimer, Kurt Koffka and Wolfgang Köhler. The author discusses gestalt theory's influence on modern art and design, describes its resemblance to Japanese-inspired theories of aesthetics and finds evidence of a mutual, if limited, interest between the gestalt psychologists and certain artists.

(Excerpt)

Gestalt psychology began in Germany in 1910. While traveling by train on vacation, a 30-year-old Czech-born psychologist named Max Wertheimer was seized by an idea when he saw flashing lights at a railroad crossing that resembled lights encircling a theater marquee. He got off the train in Frankfurt am Main, where he bought a motion picture toy called "zoetrope". When a strip of pictures is placed inside and viewed through the slits in zoetrope, a succession of stationary pictures appear to be a single, moving picture. In hotel room, he made his own picture strips, consisting not of identifiable objects, but of simple abstract lines, ranging from vertical to horizontal. By varying these elements, he was able to investigate the conditions that contribute to the illusion of motion pictures, an effect that is technically known as "apparent movement".

Years earlier, Wertheimer had studied in Prague with an Austrian philosopher named Christian von Ehrenfels, who had published a paper in 1890 entitled "On Gestalt Qualities" in which he pointed out that a melody is still recognizable when played in different keys, even though none of the notes are the same, and that abstract form attributes such as "squareness" or "angularity" can be conveyed by a wide range of specific elements. Clearly, argued Ehrenfels, if a melody and the notes that comprise it are so independent, then a whole is not simply the sum of its parts, but a synergistic "whole effect", or gestalt. Likewise, Wertheimer concluded, the effect of apparent movement is generated not so much by its individual elements as by their dynamic interrelation.

...

The three founding gestalt psychologists were separated by World War I, then reunited in 1920, when Köhler became Director of the Psychological Institute at the University of Berlin, where Wertheimer was already a faculty member. While maintaining contact with Koffka, who continued to teach near Frankfurt, Wertheimer and Köhler established a graduate program, located in the abandoned Imperial Palace, and began a research journal called *Psychologische Forschung* (Psychological Investigation). For the most part, the students did not learn by attending lectures but by actually conducting research using fellow students as subjects and by preparing articles for publication. The success of the method is evidenced by the number of teachers and students at the

Institute whose names are now familiar in psychology, including Rudolf Arnheim, Kurt Lewin, Wolfgang Metzger, Hans Wallach, Bluma Zeigarnik, Tamara Dembo, Karl Duncker, Maria Ovsiankina, Herta Kopfermann and Kurt Gottschaldt.

Koffka left Europe for the United States in 1924; Wertheimer in 1933. By the early 1930s, the psychological institute had begun to erode. When the National Socialists came to power in 1933, among their immediate menacing acts was the dismissal of Jewish university professors, from Nobel Prize scientists to graduate assistants. Rumored as being in sympathy with "the Jew Wertheimer", Köhler publicly condemned anti-Semitism and protested the dismissals in a Berlin newspaper article, the last such article allowed under the Nazis. To his surprise, he was not arrested, but the intimidation mounted, and in 1935, he too emigrated to the United States.

None of the gestalt psychologists were artists, much less designers, but early on there were signs of a mutual interest between the two disciplines. In 1927, for example, gestalt psychologist Rudolf Arnheim visited the Dessau Bauhaus, then published an article in *Die Weltbühne* praising the honesty and clarity of its building design. Soon after, gestaltist Kurt Lewin commissioned Peter Behrens(teacher of Bauhaus founder Walter Gropius)to design his home in Berlin, but, after a disagreement, Bauhaus furniture designer Marcel Breuer was asked to complete the interior. In 1929, Köhler declined a Bauhaus invitation to lecture because of a scheduling conflict, so his student Karl Duncker spoke instead. In the audience was the painter Paul Klee, who had known about Wertheimer's research as early as 1925. But other Bauhaus artists were also interested, including Wassily Kandinsky and Josef Albers, both of whom attended a series of lectures about gestalt theory by Count Karlfried von Dürckheim, a visiting psychologist from the University of Leipzig, in the winter of 1930～1931.

Surely, one of the reasons artists embraced gestalt theory is that it provided, in their minds, scientific validation of age-old principles of composition and page layout. A French byname for gestalt theory is *la psychologie de la forme*. Inadvertently, due to its emphasis on flat abstract patterns, structural economy and implicitness, gestalt theory became associated with the modernist tendency toward "aestheticism", the belief that—like music and architecture—all art is essentially abstract design and, as Ellen Lupton and J. Abbott Miller characterize it in *Design Writing Research* (1996), that "design is, at bottom, an abstract, formal activity" in which the "text(or subject matter)is secondary, added only after the mastery of form".

第2单元
Unit 2

第5课 | 概念产生的程序和方法
Lesson 5 | Procedure and methods for idea generation

词汇和短语
Words and phrases

idea generation procedure 概念产生程序

creative thinking 创造性思维

practicality 实用性

brainstorming 头脑风暴(法)

collective notebook 集体笔记本(法)

brainwriting 创意激荡(法)

mean(过去时meant) to 打算，有意要

product function 产品功能

product feature 产品特性

orthographic analysis 正交分析(法)

analogy 类推，模拟

metaphor 隐喻

cliché(法语) 谚语，陈词滥调

proverb 箴言

词汇联想与记忆
Association and memory of words

procedure *n.* 程序，手续

create *vt.* 创造，创作，引起，造成

practical	*adj.*	实际的；实践的，实用的
		应用的，有实际经验的
feature	*n.*	面貌的一部分，特征，容貌，特色，特写
	vt.	是……的特色，特写，放映
	vi.	起重要作用
collect	*v.*	收集，聚集，集中，搜集
analysis	*n.*	分析，分解

关键词汇和概念

Key words and concepts

概念就是想法、意见、主意，或一个观念。概念的产生是整个设计过程中创造性思维的核心。在概念产生过程中，不能光凭灵感，还要利用专门技术。

The idea is a thought, an opinion, a scheme, or a concept. Idea generation is the heart of creative thinking during the whole design process. Besides inspiration, we should use techniques during the process.

课文

Text

概念产生的程序和方法

Procedure and methods for idea generation

概念的产生是创造性思维的核心。所产生的概念就是创造性过程中活力的源泉。

Idea generation is the heart of creative thinking. The ideas produced are the lifeblood of the creative process.

概念产生时，首先思考的只是原则上的概念，而将其实用性留待以后考虑。第二，要从问题的正常参考框架以外去寻找想法。第三，要利用专门技术来简化问题，拓展问题，并使问题沿枝蔓引申开来。作为概念产生的程序，可以采用头脑风暴法、集体笔记本法、小组内相互启发的创意激荡等方法。

During the idea generation, firstly, think only of ideas in principle, and leave the practicalities to later. Secondly, seek ideas from outside the problem's normal frame of reference. Thirdly, use techniques for problem reduction, problem expansion and problem digression. Brainstorming, collective notebook and group-stimulated brainwriting can be used as procedures for idea generation.

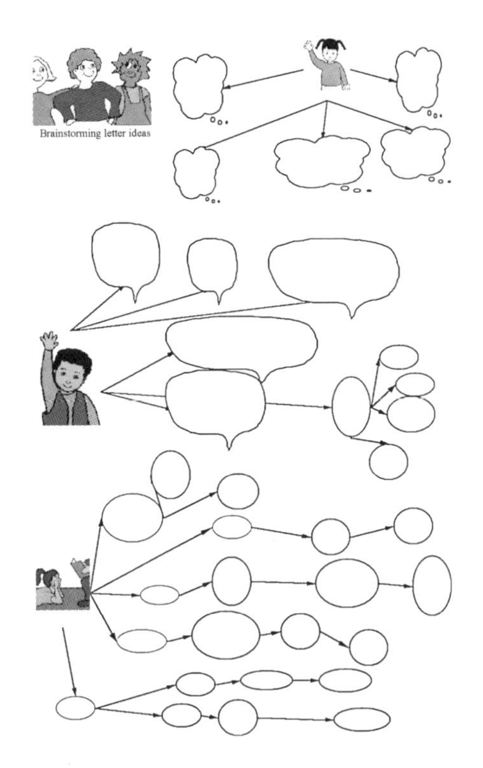
Brainstorming letter ideas

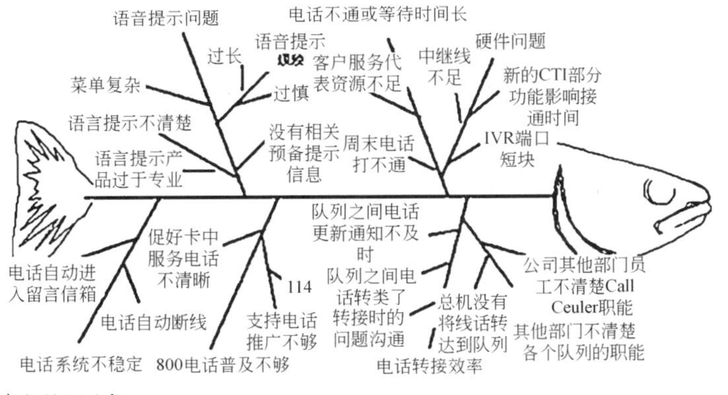
语音提示问题
过长
语音提示
过慎
菜单复杂
语言提示不清楚
语言提示产品过于专业
没有相关预备提示信息
电话不通或等待时间长
客户服务代表资源不足
中继线不足
硬件问题
新的CTI部分功能影响接通时间
周末电话打不通
IVR端口短块
促好卡中服务电话不清晰
电话自动进入留言信箱
电话自动断线
电话系统不稳定
114
支持电话推广不够
800电话普及不够
队列之间电话更新通知不及时
队列之间电话转类了转接时的问题沟通
总机没有将线话转达到队列
电话转接效率
公司其他部门员工不清楚Call Ceuler职能
其他部门不清楚各个队列的职能

| 鱼骨图分析

头脑风暴法

Brainstorming

概念产生中的关键事项是如何组织和管理该过程。很多人及多数公司认为，概念产生就是用头脑风暴法。经典形式的头脑风暴法讨论就是一小群人围桌而坐，为解决规定的问题提出新想法。头脑风暴法的关键点在于，来自每个人的想法都要为组内其他人的想法加油添彩。如果有个人提出了一条具体的想法，桌边的其他人就会在此基础上去发展和拓展它；结果使小组的产出被限制在一条思路上。讨论中有时候也会有不同的主题浮现。但研究得到的明确结论是，用经典的头脑风暴法产生的新想法的数量和质量一般要比组内每个人单独工作想出的想法少而差。这只是说要注意避免想法只汇集在有限的几个思路上去。

A key issue in idea generation is how to go about organizing and managing the process. For many people and most companies, idea generation means brainstorming. A classical brainstorming session involves a group of people sitting round a table and coming up with new ideas for solving a stated problem. A key aspect of brainstorming is that the ideas from each person are meant to fuel ideas from the other people in the group. One person will raise a particular idea and all the others round the table immediately develop and expand upon this idea, constraining the group's output to that one limited line of thinking. Occasionally, different themes will emerge during the discussion. But the clear conclusion from research is that the number and quality of new ideas produced from classical brainstorming is generally less than those of the members of the group would have produced, working individually. It just means that precautions must be taken to avoid the channeling of ideas into limited lines of thinking.

集体笔记本法

Collective notebook

用集体笔记本法时，选定一群参与者，给每人发一本空白笔记本。然后，给他们一段时间(通常是一个月)，在此期间他们在自己的笔记本上记录下各自对解决该问题的想法。在商定时间段结束时，将笔记本收集起来，并将内容摘要汇总到一个文件中。最终结果可以分发给参与者传阅，或者再召开小组会用头脑风暴法讨论提出的解决方案。这种头脑风暴法讨论就是常常能够激励出新想法或者将两个或多个想法“杂交”到一起的极好途径。

In collective notebook method, a group of participants are selected and each participant is given a blank notebook. They are then given a period of time(usually one month)in which they record their ideas for solving the problem in their individual notebook. At the end of the agreed time period, the notebooks are collected and summarized into a single document. The results are then either circulated back to the participants or a group brainstorming session is convened to discuss the proposed solution. This brainstorming session can often be an excellent way of stimulating further new ideas or hybrids of two or more individual ideas.

集体笔记本法优点明显，实际上它对参与者数量没有限制。也可以让不同地点的人参与。

The collective notebook has the great advantage that it can involve virtually any number of people. It can also involve people based at different sites.

集体笔记本法的一种变异是，在参与者记录了自己想法2～3周后互相交换笔记本。

A variation on the collective notebook procedure is to have participants exchange notebooks after two or three weeks of recording their own ideas.

此外，有许多专门技术可用于概念产生。例如：产品功能分析法，产品特性分析法，SCAMPER法，正交分析法，强制配合表法，产品“杂交”法，类比和隐喻法，谚语和箴言法等。

Besides, there are many special techniques which can be used for idea generation. For example, product function analysis, product feature analysis, SCAMPER, orthographic analysis, force fit tables, product hybridization, analogies and metaphors, clichés and proverbs, and so on.

拓展阅读

Extend and reading

The loop chair

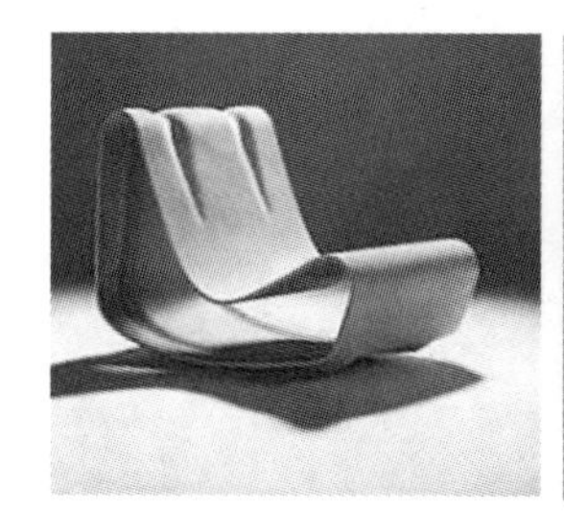

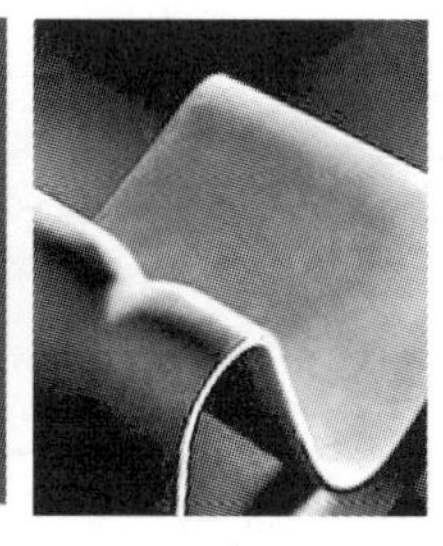
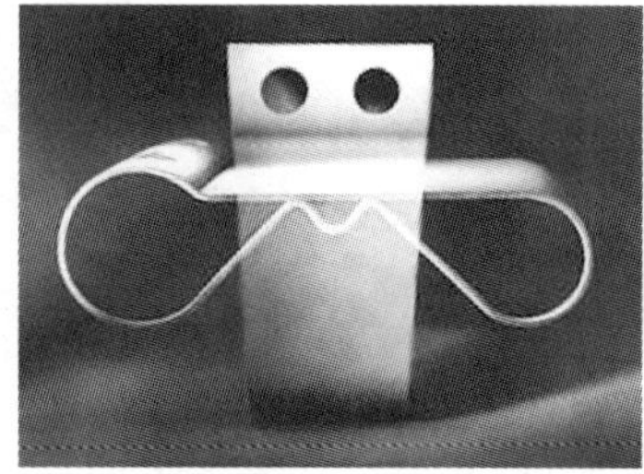

These two items are a fascinating mix of indoor and outdoor furniture. Both pieces are handmade in Switzerland of Eternit: a light gray, asbestos free, cement/fiber bond. The chair, which would look amazing at any pool side or in a large loft space, is actually a rocking chair. It is lightweight and easy to move around, making it a perfect lounge chair. The loop table, with 2 recessed drink holders, fits perfectly into the base of the loop chair for storage. They are extremely comfortable, and a great addition to the velocity line of products. We love the look and the feel of this item, and thought it would be perfect to add a chair to our inventory that could be put at the pool side for those long summer evenings spent sipping on your favorite cocktail.

The legendary Eternit Beach chair designed by Willy Guhl in 1954 is a classic of modern furniture design. In the collections of the Philadelphia Museum of Art, the Vitra Design Museum and the Boyd Collection, the 1998 redesign by Willy Guhl himself demonstrates again the designer's credo of "achieving the optimum with minimum effort".

The loop chair is unsurpassed as a self-contained and unsupported fibre/cement bond chair. Whether used indoors or outdoors, this chair is a sculpture and work of art. The material Eternit, as asbestos free cement/fibre bond, has a smooth, warm surface and is almost indestrucible. It is handmade in Switzerland.

第6课 | 概念设计
Lesson 6 | Concept design

词汇和短语
Words and phrases

core benefit proposition　核心利益主张

arm with　用……武装

functional principle　功能原则

styling principle　造型原则

intuition　直觉

imagination　想象力

out of the blue　出乎意外

inspiration　灵感

problem gap　问题间隙

problem boundary　问题边界

existing solution　现有方案

词汇联想与记忆
Association and memory of words

concept　*n.*　观念，概念

core　*n.*　果核，中心，核心

benefit　*n.*　利益，好处

vt.　有益于，有助于

vi.　受益

image	*n.*	图像，肖像，偶像，典型
	vt.	想象反映，象征
boundary	*n.*	边界，分界线
solution	*n.*	解答；溶解，溶液；解决方案

关键词汇和概念
Key words and concepts

概念设计旨在为新产品设定设计原则。成功的概念设计有两个简单的秘诀：产生的概念越多越好，然后挑选出最佳方案。

Concept design aims to produce design principles for the new product. There are two simple secrets of successful concept design: generate concepts as more as possible, and select the best.

课文
Text

概念设计
Concept design

概念设计旨在为新产品设定设计原则。这些原则应该足以满足消费者的要求，并使该产品与市场上的其他产品不同。明确地讲，概念设计要表达一个新产品如何传递它的核心利益主张。因此，有效的概念设计的先决条件是有确定的核心利益主张，并对客户需求以及竞争产品有充分的理解。据此，概念设计就是要着手设定一整套产品将如何工作的功能原则以及一整套产品外观将看起来如何的造型原则。

Concept design aims to produce design principles for the new product. These should be sufficient to satisfy customer's requirements and differentiate the products from others on the market. Specifically, concept design should show how the new product will deliver its core benefit proposition. Prerequisites for effective concept design are, therefore, a defined core benefit proposition and a good understanding of both customer needs and competing products. Armed with this information, concept design sets about producing a set of functional principles for how the product will work and a set of styling principles for how the product will look.

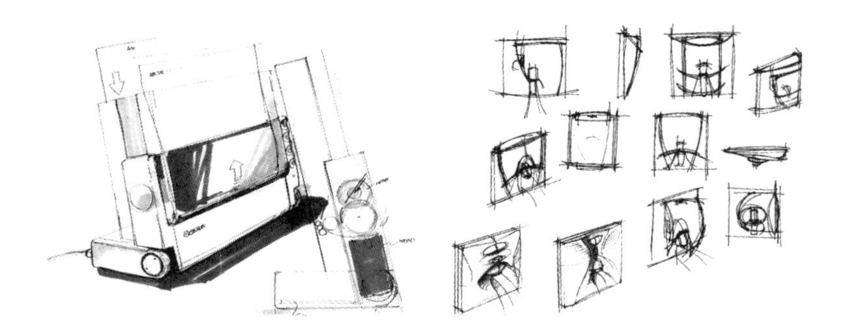

概念设计过程

The concept design process

成功的概念设计有两个简单的秘诀。首先要产生很多很多的概念，然后要挑选出最佳方案。概念设计通常是产品开发中最需要非凡创造性的一步。正是在这个阶段里萌发了许多发明。

There are two simple secrets of successful concept design. Firstly, generate lots and lots of concepts and secondly select the best. Concept design is the stage of product development which usually demands the greatest creativity. It is at this stage that inventions are invented.

确定概念设计的目标和范围

Establishing the aims and the scope of concept design

对不同的产品而言，概念设计的目标和范围可以大不相同。例如，若一个确定的机会要求很快生产出一种现有产品的低成本型新产品时，那么对该产品提出一套全新的工作原理可能是达不到预期目标的。反过来说，如果目前市场上所有的产品都不能满足一个确定的消费者需求，那么对设计原则从基本原理上进行再思考可能就是必要的了。

The aims and scopes for concept design differ greatly for different products. If, for example, the identified opportunity is to rapidly produce a reduced-cost version of an existing product, coming up with a whole new set of working principles for the product may be counter-productive. If, on the other hand, all products currently on the market fail to satisfy an identified customer need, then a fundamental re-think of design principles may be essential.

确定概念设计要求的目标和范围可以从确定问题间隙着手。所谓问题间隙是现有解决办法与问题目标之间的空间，这个空间又被问题边界所限制。

Establishing the aims and scopes of the concept design required can be tackled by plotting the problem gap. The problem gap is the space between existing solutions and the problem goal, and is limited by the problem boundaries.

问题间隙分析的下一步是探究问题边界。会有什么样的限制加到可能产生的潜在概念范围上呢？这是些以商业上现实的方式影响到新产品核心利益的设计限制。

The next stage of problem gap analysis is to explore the problem boundaries. What constraints are appropriate to be imposed on the range of potential concept which could be generated? These are the design constraints imposed on how the core benefit of the new product is to be delivered in a commercially realistic way.

一个可能的典型限制是新产品必须能制造出来。另一个限制可能是，新产品要适合通过现有的销售渠道或者明确给出的新的潜在销售渠道进行销售。再一个限制可能是新产品中有特别的新元件、新技术等。

A typical constraint might require that the new product can be manufactured. Another might be that the new product is suitable for sale by existing sales outlets, or alternatively, that it specifically gives access to new potential sales outlets. Another could be that the new product incorporates a specific new component or technology.

要牢牢记住，概念设计从根本上讲是一个创造性过程，只需要设定整个产品的功能和造型的原则。重要的是要尽量为概念的产生保留尽可能多的可能选择，在想法和主意产生之后再用严格的概念选择程序来选出最佳方案。概念设计通常认为是设计过程中的创造性核心。

Remember that concept design is a fundamentally creative process and only goes as far as proposing functional and styling principles for the entire product. It is important to try to keep as many options open as possible for concept generation and then impose rigorous concept selection procedures to select the best one once the ideas have been generated. Concept design is usually considered to be the creative heart of the design process.

好的概念设计要求使用直觉、想象力和逻辑性，对现已确定的问题提出创造性的解决方案。

Good concept design requires the use of intuition, imagination and logic to come up with creative solutions to the now well-defined problem.

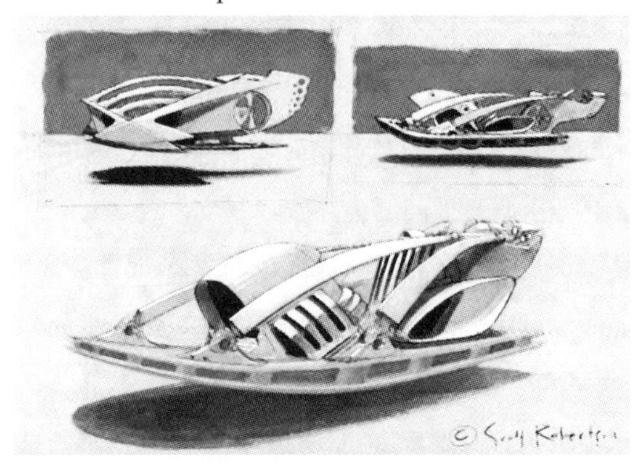

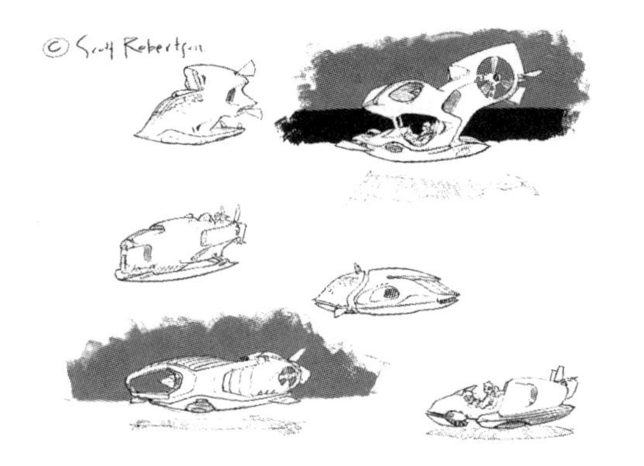

拓展阅读
Extend and reading

Item：softwall 48"

designer：Molo

Perfect for a playroom! The paper softwall is a beautiful，lightweight，freestanding wall that can be arranged into almost any shape，or easily compressed into a compact sheaf and stored away. Softwall dampens sound and can both absorb and transmit light. The paper softwall is made from 400 layers of honeycombed translucent white，fire-retardant paper，bounded by natural wool felt ends. The thick felt ends fold to create handles when the wall is open，and form a casing when the wall is compressed. Paper softwall is modular，as the felt ends have velcro fasteners which can link walls together. The paper softwall is delicate，yet its honeycomb design makes it surprisingly resilient to normal handling.

Softwall can be opened by one person，although it is initially easier to open with two people. Softwall moves best when it can glide on a smooth，clean，dry surface. When not in use，softwall can be stored away flat，or folded in half.

Softwall is not designed to support weight，or to be altered. The paper softwall can be dusted，or lightly vacuumed with a brush handle，but should not be allowed to get wet. The paper body of softwall is treated with a fire retardant and will not maintain a flame on its own；nevertheless，the paper softwall should not be placed near an open flame or heat source. Should a tear in the paper ever occur, paper softwall can be mended to conceal and protect the affected area. Softwall is also reversible, as it is identical on either side.

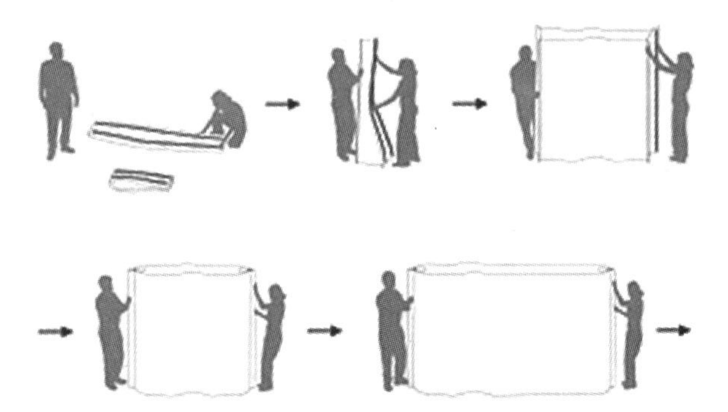

At 4 feet tall, this softwall can be used to dynamically partition space, or create the sense of separate space while only partially impeding sight lines, whether standing or sitting. This softwall is 12" thick(30cm), and can be expanded from a mere 1.5"(3cm)to over 25'(7.5m). It is available in white paper with medium grey natural wool felt ends, and weighs just 10lbs (4.5kg).

Item: Uten.Silo

designer: dorothee becker and ingo maurer

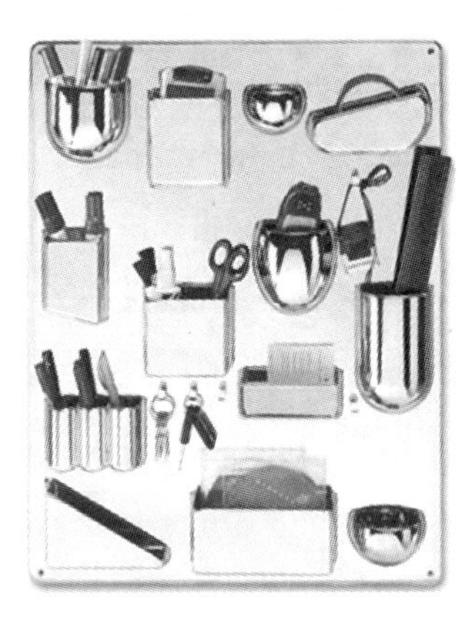

Uten.Silo is one of the best-known plastic designs of the late 1960s. Today, this colorful wall tidy is a veritable design icon and a true-to-the-original version of the product has now been reissued by Vitra Design Museum.

At the end of the 1960s plastic was on its way to becoming one of the dominant design materials. In Italy designers such as Joe Colombo and Vico Magistretti designed bright and cheerful furniture, luminaires and consumer goods for innovative manufacturers such as Artemide and Kartell. With

Bofinger Chair and Panton Chair the first seats made entirely of plastic appeared. At the same time, in Munich, Ingo Maurer, who had already attracted attention with his "Bulb" luminaire, launched a plastic wall tidy which created quite a stir—Uten.Silo.

Uten.Silo was designed by Maurer's wife, Dorothee Becker. With its differently-shaped and sized pockets, its metal hooks and clips Uten.Silo organizes offices, kitchens, bathrooms and children's rooms. The tension between industrial precision and playful variety, between logical organization and humorous design makes Uten.Silo a highly functional design which also allows plastic to be put to a sensible use. The main body was Made of ABS plastic, with metal hooks according to an original model.

第7课 | 概念选择
Lesson 7 | Concept selection

词汇和短语
Words and phrases

concept and idea selection　概念和方案选择

selection criteria　选择标准

concept selection matrix　概念选择矩阵

rank　排队，排序

reference concept　参考概念，参照概念

hybridization　杂交，杂种培植

relative merit　相对价值

problem definition　问题定义

dot sticking(technique)　贴点法

sticky dot　带背胶的圆点贴纸

marketing staff　市场营销人员

词汇联想与记忆
Association and memory of words

idea　*n.*　想法；主意，思想，观念；概念

criterion　*n.*　(批评判断的)标准，规范

reference　*n.*　提及，涉及，参考，参考书目；证明书(人)，介绍信(人)

relative　*n.*　亲戚，关系词；相关物

adj. 有关系的，相对的

staff *n.* 棒，杖；杆，支柱，全体职员

vt. 供给人员；充当职员

关键词汇和概念

Key words and concepts

概念选择是个概念筛选过程，经常需要大量的创造性。正是在这个阶段中，想法需要扩展、开发，甚至“杂交”以得到越来越理想的解决办法。

Concept selection is an idea screening process that often requires a great deal of creative. It is at this stage that ideas must be expanded, developed and even hybridized to get closer and closer to an ideal solution.

课文

Text

概念选择

Concept selection

解决问题，甚至整个产品设计过程的一个基本特点就是要考虑所有可能的解决办法，并从中选出最佳解决办法。通常会误解，认为解决问题的创造性部分只到概念产生为止，而选择方案只是不用脑子的例行程序而已。事实上，在筛选概念的过程中经常需要大量的创造性。其实，正是在这个阶段中，想法需要扩展、开发，甚至“杂交”以得到越来越理想的解决办法。

An essential feature of problem solving, and indeed the whole of product design, is to think of all possible solutions and pick up the best. A common misunderstanding is that the creative part of problem solving ends with idea generation and that idea selection is rather mindless routine procedure. In fact, a great deal of creative is often required during idea screening. It is at this stage that ideas must be expanded, developed and even hybridized to get closer and closer to an ideal solution.

比起概念的产生来说，概念和想法的选择是更为严格、系统和训练有素的过程。概念和想法选择的目标是，从已经创造出的广泛而富有想象的概念和想法中，识别出能解决最初抓住的问题的最好方案来。

Concept and idea selection is a more rigorous, systematic and disciplined procedure than idea generation. Concept and idea selection aims to identify from the wide and imaginative range of concepts and ideas created, those which best solve the problem originally tackled.

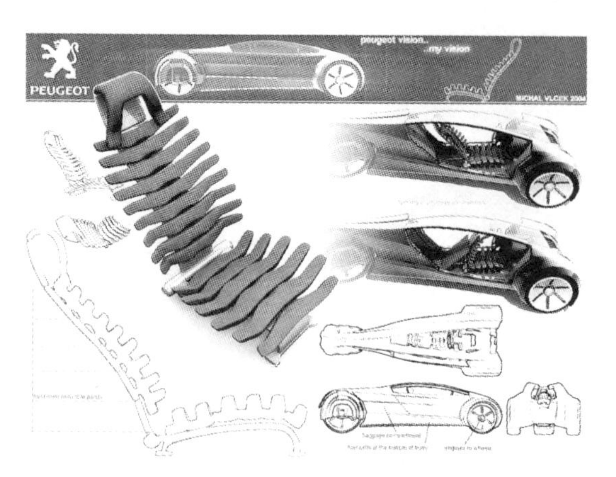

概念选择的第一轮是，对照一系列的选择标准将概念排序。这件事可以用概念选择矩阵来做：概念沿矩阵的一个轴列出，而选择标准沿另一个轴列出。为简化排序程序，每个概念仅与参考概念进行比较，判定为“更好”(打+1分)、“更差”(打−1分)或“相同”(打0分)。参考概念对于计划中的新产品应该是目前最好的竞争者。排序过程的结果是个数字，表示每个概念的相对价值(正数表示此概念整体上优于参考概念，而负数表示整体上比参考概念差)。这样排列可使注意力集中到较优的概念上。

The first round of concept selection ranks the concepts in relation to a series of selection criteria. This is done by means of a concept selection matrix in which the concepts are arranged along the one axis of the matrix and selection criteria along the other. To make the ranking procedure simple, each concept is judged “better than” (scored as +1), “worse than” (scored as–1)or “the same as” (scored as 0)a reference concept. The reference concept should be the best current competitor to the proposed new product. The outcome of the ranking process will be a single number expressing the relative merit of each concept(a positive number indicates that the concept better overall than the reference concept, a negative number indicates worse overall than the reference). From these ranks, attention focuses on the better concept.

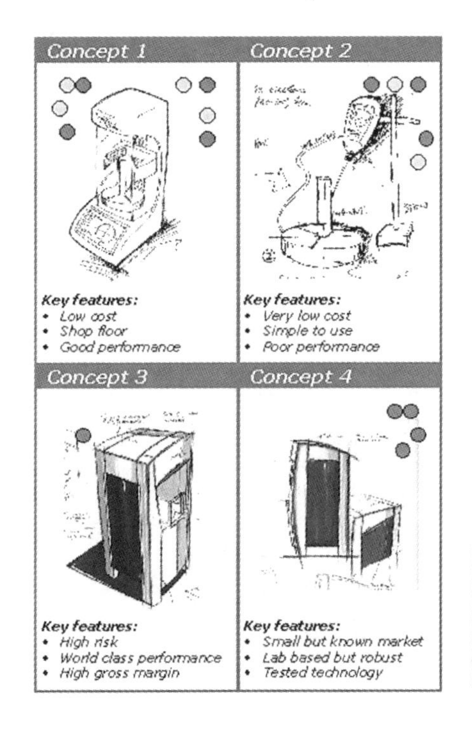

给若干位评判人发一定量带背胶的贴点，用它贴在带有方案的纸上，让他们用这种办法来选出潜在的解决办法。
Several judges are given a number of sticky dots which they use to vote for potential solutions by sticking them to the sheets of paper describing the solution.

还有第二种方法可使用于概念、方案和解决办法的选择——贴点法。给若干位评判人发一定量带背胶的贴点，用它贴在写有方案的纸上，让他们用这种办法来选出潜在的解决办法。用贴点法投票的优点是可以将不同颜色的贴点发给公司内不同职能的人员。这样就可能显现出营销人员偏爱某些方案，而生产工程师可能偏爱另一些方案。

There is second method—dot sticking, which can be used for selecting concepts, ideas and solutions. Several judges are given a number of sticky dots which they use to vote for potential solutions by sticking them to the sheets of paper describing the solution. An advantage to voting with dot sticking is that different color can be given to people with different functions within the company. Thus it may emerge that some ideas are preferred by marketing staff, whereas others are preferred by the production engineers.

贴点法技术的关键部分是投票程序后的讨论。这样可以确定，为什么某些方案胜过其他方案。再进行简短的头脑风暴法讨论，可以改进较少受欢迎的方案使它们趋于最佳。另外，好的方案也要加进差方案中的好特性来予以发展。时刻记住，扔掉一个比其他想法有某些优点的概念可能是一个代价高昂的错误。

A key part of the dot sticking technique is the discussion that accompanies the voting procedure. This should establish why some ideas are preferred over others. Short brainstorming session can then try to improve the less favored ideas to make them match the best. In addition, the good ideas should be developed to try to incorporate any good features in the poor ideas. Always remember that throwing away an idea with any advantage over the other ideas can be an expensive mistake.

拓展阅读

Extend and reading

Dot sticking

It's a good way to engage a wide number of stakeholders in concept selection. Simple and easy to administer, but does not necessarily provide rich feedback behind the reasons or motivations for choice.

The dot sticking approach is a good selection tool, when there are a wide array of potential or competing ideas as well as a large number of stakeholders. By allocating each stakeholder with a number of sticky "dots", they can allocate one, some or all of them to their preferred choice.

Approach

1. Define concepts

Ensure that each concept is presented to a similar level of detail, sufficient to enable suitably qualified stakeholders to make an informed selection.

2. Determine the stakeholders

Who is to make the choice? Is it to be just a representative sample of internal stakeholders, or will external stakeholders also be involved?

3. Allocate "dots"

Allocate each stakeholder with a number of "dots". Depending upon the number of concepts, this can range from 3 to 5. Different coloured dots can be used for a variety of purposes. Each person could have a different coloured dot to assign to his or her most preferred choice—this can be useful in a tiebreak. Alternatively, different coloured dots can be given to representatives of different functions or customers from different market segments, to determine whether there is any bias depending upon perspective.

4. Vote and filter

Following the initial voting, it can be useful to filter out the results and have a second pass, this time only keeping in the strongest contenders.

5. Capture likes and dislikes

It is important to capture some of the reasoning behind the selection, so it can be useful to encourage participants to note likes and dislikes on post-it notes and also attach these to the different solutions. This will help generate understanding as to why some ideas are preferred over others. It also enables strong features of rejected ideas to be kept.

Example: Concept and idea selection

Concept and idea selection aims to identify from the wide and imaginative range of concept and ideas created, those which best solve the problem originally tackled.

Was each idea discussed and weighted up against supporting and refusing evidence?	5 4 3 2 1	1 2 3 4 5	Did the group deal well with people whose ideas were being criticized or rejected?
Was the group systematic in its use of selection criteria?	5 4 3 2 1	1 2 3 4 5	Did the group concentrate on selecting the best idea rather than rejecting the poor ideas?
How well did the group modify and combine the initial ideas?	5 4 3 2 1	1 2 3 4 5	Were differences of opinion negotiated to a point of mutual satisfaction?
Was one idea(or set of ideas)finally selected for more thorough exploration and evaluation?	5 4 3 2 1	1 2 3 4 5	Was the solution chosen by consensus and if not, was the extent of the agreement within a group established?
Were the rules of the idea generation session reviewed and agreed at the start?	5 4 3 2 1	1 2 3 4 5	Were all ideas recognized and welcomed, regardless of their content?
Was everyone's idea generating capacity exhausted at the end?	5 4 3 2 1	1 2 3 4 5	Were less forthcoming group members encouraged?
Once all ideas were generated, were they reviewed by the group for clarification, elaboration or addition?	5 4 3 2 1	1 2 3 4 5	Was criticism tactfully discouraged and evaluation effectively postponed?
Were the ideas clustered into sets with common features or attributes?	5 4 3 2 1	1 2 3 4 5	Was any one member prevented from dominating the discussion or imposing their ideas on the group?
Was a summary list of the most innovative, feasible or interesting ideas produced?	5 4 3 2 1	1 2 3 4 5	Was the ideas finally presented or posted for all to see?

第8课 | 阿莱西公司
Lesson 8 | Alessi

词汇和短语
Words and phrases

immemorial　古老的，远古的

forebear　祖宗，祖先

pewter　锡镴，白镴

launch　投放市场；发射

helm　舵

conquer　征服，占据

vacate　腾出(空间，职位)，退位

chrome　铬，铬合金；镀铬

silver-plated　镀银的，包银的

alloy　合金

authority　权威，威信，权力

catering　公共饮食业，给养

big seller　大卖家；畅销品

utopian　乌托邦的，理想化的

manifesto　宣言，声明

champion　冠军；支持，拥护

veritable　真正的

conceive　构思，设想

pursue　追赶，追踪，持续

anthropology 人类学

metaproject 变形设计，变形规划

typology 形体学，象征学；血型学

stimulus(复数stimuli) 刺激，激励

elaborate 详细阐述，精心描述

coordinate 调整，整理，协调

conceptualize 使有概念，概念形成

fiction 虚构，小说

ritual 仪式(的)，(宗教)典礼(的)

rigour 严格，严峻

severity 严肃，严格

词汇联想与记忆
Association and memory of words

elaborate *adj.* 精心制作的，详细阐述的
vt. 精心制作，详细阐述
v. 详细描述

coordinate *n.* 同等物，坐标(用复数)
adj. 同等的，并列的
vt. 调整，整理

inspirit *v.* 激励

plated *adj.* 镀金的，装甲的

project *n.* 计划，方案；事业，企业；工程
v. 设计，投射；放映，发射

stimulus *n.* 刺激物，促进因素；刺激

关键词汇和概念
Key words and concepts

阿莱西公司是一家历史悠久的意大利公司，它出售的商品以具有真正的艺术性而著称。这得益于其对沟通和人类学领域以及艺术和营销所进行的研究。

Alessi is an Italian company with long history, and is well-known for offering veritable artistic commodities. It comes from researching the fields of communication and anthropology, as well as art and marketing.

课文
Text

阿莱西公司
Alessi
第一部分
Part Ⅰ

从远古时代起，阿莱西家族就已经在奥尔塔湖边稳固地建立了。我们的祖先远赴德国学习锡镴制作生意。其中一些人回到家乡并开办了最早的手工艺作坊。这是奥梅格纳地方的金属家用物品制造者的开端，也是今天欧洲此类物品生产最具活力的中心之一。1921年阿莱西公司成立了。

Since immemorial time the Alessi family has been firmly established on Lake Orta. Our forebears went as far as Germany to learn the trade of pewter-maker. Some of them returned home and opened the first craft workshops. Such were the beginnings of the Omegna makers of metal household objects, today one of the most dynamic centers in Europe for the production of such items. In 1921, Alessi was founded.

我父亲卡洛受过工业设计师的培训，他在很年轻时就加入了阿莱西公司。1945年他把他的最新设计，一个称为Bombé的咖啡和茶具投放了市场。20世纪50年代，他从我爷爷那里接掌了公司。第二次世界大战后，我父亲很早就意识到不锈钢将要占领镀铬和包银合金所让出的市场。

My father, Carlo, trained as an industrial designer, joined the company when still very young. In 1945, he launched his last project, called the Bombé coffee and tea sets. In the 1950s he took over from my grandpa at the helm of the company. After the World War Ⅱ, my father had realized so early that stainless steel was set to conquer the space vacated by chromed metals and silver-plated alloys.

比我父亲小11岁的叔叔依陀尔是金属冷压方面的权威。1955年，他开创了阿莱西与外界设计师的合作之路，他与许多建筑师的合作造就了范围广泛的产品，特别是餐饮业用品，其中许多至今仍然是畅销品。

Uncle Ettore, eleven years younger than my father was the great authority on cold pressing of metals. In 1955, he opened Alessi up to collaboration with external designers. His work with many architects produced several ranges of products, particularly for the catering trade, many of which are still big sellers.

1970年7月，我从法律专业毕业后就正式在阿莱西开始工作。我以一种强烈的乌托邦观点发展了我的文化理论宣言，主张商业文明要用低廉的价格向消费大众提供真正有艺术性的物品。

Officially my career at Alessi began in July 1970, the day after I graduated in law. With a strongly utopian view, I developed my cultural-theoretic manifesto championing a new commercial civilization offering the consuming masses veritable artistic items at low prices.

——阿尔伯特·阿莱西

——Alberto Alessi

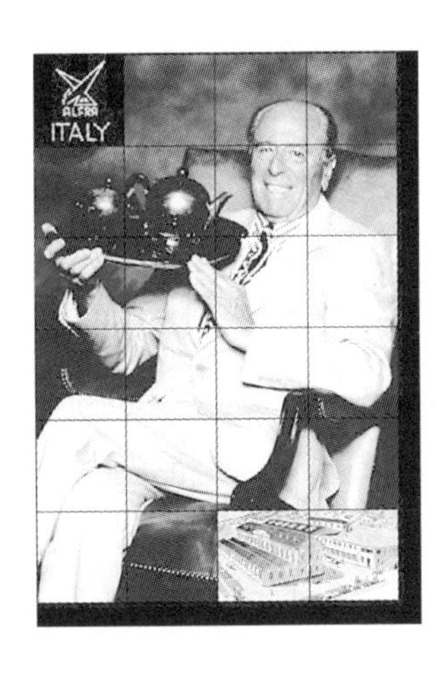

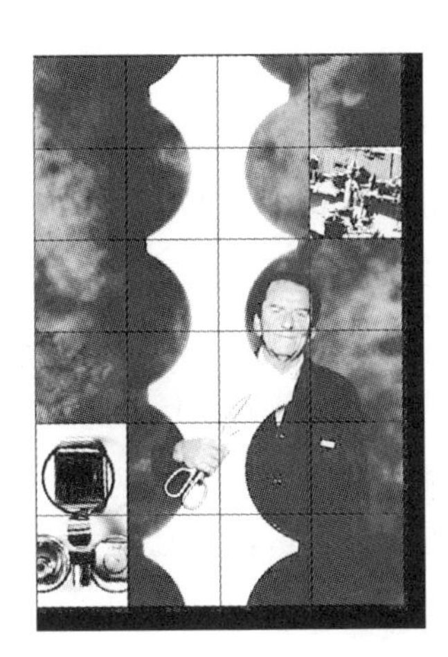

第二部分

Part II

阿莱西中央设计工作室(CSA)是20世纪80年代末从阿尔伯特·阿莱西、亚历山德罗·曼蒂里和劳拉·普里诺罗的一个想法所构思建立起来的；这三个人提出了创办一个研究中心来整合和革新阿莱西已经确立的项目体系的最初构想。

Centre Studi Alessi(Alessi Central Studio, CSA)was conceived and established at the end of the 1980s from an idea of Alberto Alessi, Alessandro Mendini and Laura Polinoro, who were called to develop the initial idea of creating a research center to integrate and innovate Alessi's established project system.

从1990年起，CSA持续要求探索新的设计之路，以使年轻设计师在沟通和人类学领域以及艺术和营销方面进行研究，以寻找新的美学灵感。

Since 1990，CSA was pursuing the need to explorenew ways to design in order to make young designers researching the fields of communication and anthropology，as well as art and marketing，to find inspirations for a new aesthetic.

变形设计是一种新的设计方法，它被定义为被设计物体的体形学。它收集公司的需求并给出美学准则以及文化和视觉对设计师的刺激。

The metaproject is one of the new ways of design and is defined as the typologies of objects to design.It collects the company's needs，and gives the aesthetical criteria as well as the cultural and visual stimuli to the designers.

通过这个平台变形设计得到了详细的阐述，想法也在选定的设计师与阿莱西的技术队伍之间交流。CSA整合这些交流，绘制出首批设计草图，并挑选了以生产为目的需要进一步开发的一批项目。由劳拉·普里诺罗为CSA提出并予以协调的变形设计这个概念的头两次运作是："记忆容器"和"虚构家庭(FFF)"。"记忆容器"的灵感来自于对用于食物祭奠的物体(盘、碗等)有关的个人和集体追忆的研究。"虚构家庭"的构思来自于给冷冰冰、严肃的不锈钢物体加上有趣而温暖的感觉的一个构想，以探索对于阿莱西世界而言新的基础并由此开始使用塑料。

Through the workshop，the metaproject is elaborated and ideas exchanged with the selected designers and the technical team of Alessi. The CSA coordinates this exchange，addresses the first draft of designs and selects the group of projects that will be further developed for production purposes. The first two operations—metaprojects conceptualised and coordinated by Laura Polinoro for CSA were："Memory Containers" and "Family Follows Fiction—FFF". "Memory Containers" was inspired by a research on individual and collective memory related to objects intended for the ritual of the offer of food (trays，bowls etc.). "FFF" was conceived by the will of adding a playful and warm feel to the rigour and severity of the stainless steel objects，exploring grounds new to the world of Alessi and starting to use plastic.

CSA在许多大学和研究中心举办了一系列的专题研讨会议，如伦敦皇家艺术学院、米兰理工大学和朵玛斯学院、赫尔辛基艺术设计大学、东京ICS、佛罗伦萨大学和首尔的韩国设计促进院。

A series of workshops and conferences are held by CSA among some universities and research centers，such as Royal College of Art，London，Politecnico and Domus Academy of Milan，UIAH of Helsinki，ICS of Tokyo，The University of Florence and KIDP of Seul.

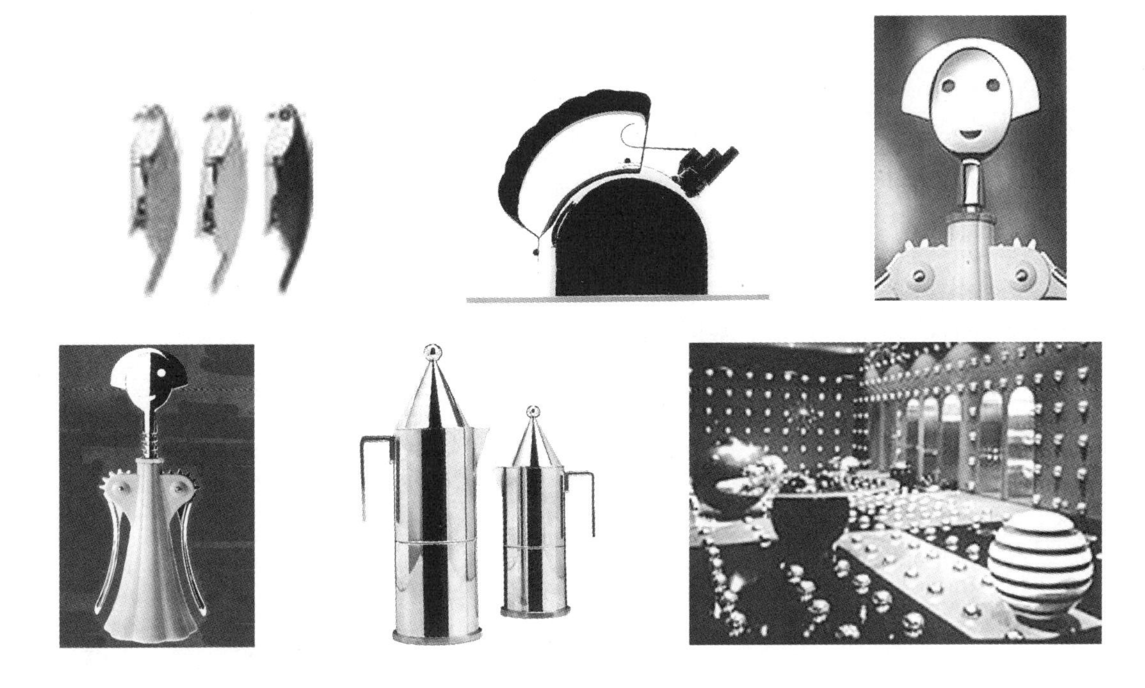

拓展阅读

Extend and reading

Alessi's designer：Alessandro Mendini

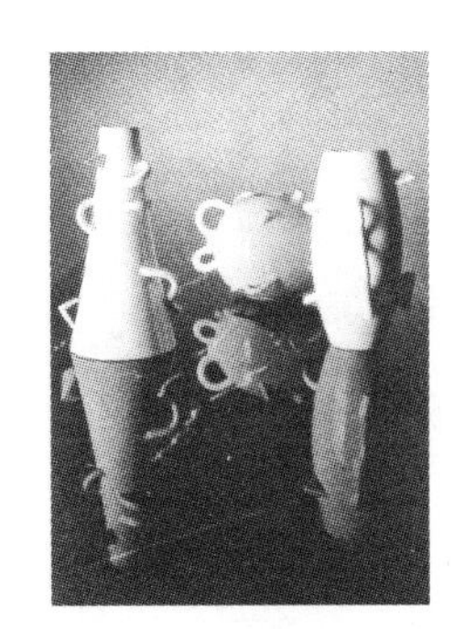

Born in Milan in 1931, former director of Domus, winner of a Compasso d' Oro, he is a designer, architect and image consultant for Philips and other companies.

"He has been something of a mentor for me. When people ask me what Mendini does for us, what his role is, I just smile. Alessandro Mendini's Consulting work covers such a wide range of aspects that his position cannot really be described or even understood by an outsider."

"As a designer, he continues to come up with objects for us, which often are to be found in the most difficult and exciting areas of our catalogues.

As an architect, he has designed for us the Alessi Museum and so on. As a design manager, he is responsible for conceiving and overseeing some of our renowned design projects. As a journalist, he has written books for us. As a consultant, he has introduced me to a large number of designers who have worked for us. As a close friend, he is aware of my most intimate problems and aspirations ... over the years we have achieved a working affinity, a relationship that is almost telepathic..."

Questions about myself

The main traits of my character.

Willingness.

A quality I desire in a man.

Complexity.

A quality I desire in a woman.

Complexity.

What I appreciate most among my friends.

Niceness.

My principal defect.

Narcissism.

My favourite occupation.

Thinking

My dream of felicity.

To not think.

What would for me be the biggest misfortune.

The end of the world.

Whom I would like to be.

A saint.

The country were I would like to live.

Where I live.

The colour I prefer.

Pink.

The flower I love.

The rose.

The bird I prefer.

The woodpecker.

My favourite authors.

Nietzsche.

My favourite poets.

Tangore.

My heroes in fiction.

Mickey Mouse.

My heroines in fiction.

Alice.

My favourite composers.

Schubert.

My favourite artists.

Savinio.

My heroes in real life.

Gregory Peck.

My heroines in history.

Queen Victoria of England.

My favourite names.

The names of the apostles.

What I hate most.

Violence.

The historic characters I dislike most.

Dictators.

The reformation I appreciate the most.

Buddhism.

Nature's gift I would like to have.

Ubiquity.

How I would like to die.

In my bed.

My soul's present condition.

Slightly anxious.

The faults I can bear.

All of them.

My motto.

Uncertainty.

第3单元
Unit 3

第9课 | 系统设计的基本准则
Lesson 9 | Ground-rules for systematic design

词汇和短语
Words and phrases

ground-rules　基本准则

evil　罪恶，不幸，诽谤

clairvoyance　洞察力，千里眼

compatibility with　与……的兼容性

suitability for　与……的适应(适宜)性

sales and distribution channels　销售分配渠道

conformance with　与……的一致性

statutory　法律法规的

adage　格言，谚语

go off　偏离

sound out　试探，调查

sounding　收集意见，调查(结果)

come up with　提出

词汇联想与记忆
Association and memory of words

ground　*n.*　地面，土地，场所，范围

adj.　土地的，地面上的

vt.　把……放在地上，打基础

compatibility	*n.*	[计]兼容性
suitability	*n.*	合适，适当，相配，适宜性
distribution	*n.*	分配，分发；销售，分布状态区分，分类；发送，发行
channel	*n.*	通道，通路
sound	*n.*	声音，噪声，吵闹，海峡
	adj.	可靠的，合理的，有效彻底的

关键词汇和概念

Key words and concepts

系统设计就是以系统的观点来处理设计事务的设计观念和方法，而系统是许多事情或角色以有规律的关系在一起工作形成的一个组合。

Systematic design is a design idea and methodology of design to govern all design activities in the view point of system, which is a group of things or parts working together in a regular relation.

课文

Text

系统设计的基本准则

Ground-rules for systematic design

系统设计的三个基本准则的第一条是：不要看弊病！在新产品开发的词典中，一开始就看到方案的弊病就意味着定位了一个在市场上将要失败的新产品，这对新产品的开发是致命的。创造一个不会失败的产品这个任务太重要了，因此不能单凭洞察力来完成，而必须要用系统的手段来处理解决。对新产品设定一个明确的和现实的目标，为产品的必定成功提供了一种先见之明。最重要的目标是客户需求或希望要的东西。其他重要目标还有：与生产厂的能力和设备的兼容性，与目标市场和分销渠道的适应性，以及与相关法令或工业标准的一致性。设计者如果不能定出目标，那才是真正的“看不到弊病”。正如古老的格言所说：“如果你不知道去向何方，那么条条大路都是对的”。

The first of three ground-rules for systematic design is to see no evil! Seeing evil, which in the new product development dictionary means spotting the new products which will fail in the marketplace, is vital in new product development. To create products which will not fail is far too important a task to be left to clairvoyance and must, therefore, be tackled systematically. Setting clear and realistic targets for a new product provides the vision of what that product must achieve to be successful. The most important targets are those demanded or wished for by customers. Other important targets include compatibility with the skills and facilities of the manufacturer, suitability for the intended marketing, sales and distribution channels and conformance with relevant statutory or industry standards. Designers who fail to set targets will indeed "see no evil". As the old adage says "if you do not know where you are going then every road is the right one".

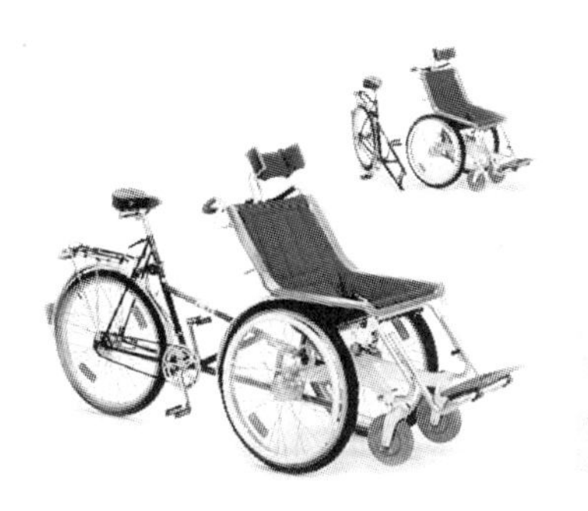

第二条基本准则是不要听弊病！如果在整个开发过程中对进程不予监控，则为产品设定目标就没有多大意义。判定从什么时候起产品偏离了正常路线的唯一方法是，对照已设定的目标，以一定的间隔时间进行诊断并检查这些结果。设计师如果对于新产品出差错的议论充耳不闻的话，那他们只能指望对新产品上市时的失败“声响”也充耳不闻。

The second ground-rule is to hear no evil! Setting targets for the product is of little value unless its progress throughout the development process is monitored. Sounding out the product at periodic intervals and checking these soundings against targets is the only way to spot when the product has started to go off course. Designers who are deaf to the tell-tale signals of a new product going wrong can only hope they are enough to miss the crash when the product is launched!

设计师如果对新产品出现差错的议论充耳不闻的话，
那他们只能指望对新产品上市时的失败“声响”也充耳不闻。
Designers who are deaf to the tell-tale signals of a new product going wrong can only hope they are enough to miss the crash when the product is launched.

第三是不要说弊病！创造性的自由表达是设计的核心。用系统设计方法去撞击沉默不语的创造性，是个广为流传的神话。但确实再没有比这更真实的事情了。正如托马斯·爱迪生曾说过的，创造力是1%的灵感加上99%的辛劳。这份辛劳之汗是在你武装头脑时流的——建造基础，使创造力的大厦可以在这个基础上建起来。对真实、伟大发现的历史记载都乐于关注那些造就突破的最终的直觉飞跃；而很少提及在研究问题、反复思考现有解决办法并探索所有最终被证明是毫无用处的那些办法上所花费的漫长岁月。在许多此类案例中，正是这些错误的解决方案痛苦地、一步一步地把我们引向了突破。新产品开发也应该是这种情况。为了发现那些成功的创意，就必须给创造力以自由、容忍提出不成功的想法。事实上，被驳回的想法的数量和质量正是一个人产生概念能力的最好衡量。说得更极端一点，对一个问题的各种可能解决方案提出得越多，你就越可能找到最好的方案。因此，自由地表达创造性想法，包括那些最后被证实是一钱不值的想法，在新产品开发中实在是个优点。

The third is to speak no evil! Freedom of creative expression is at the heart of design. And it is one of the most widely held myths that the use of systematic design methods strikes that creativity dumb. Nothing could be further from the truth. Creativity, as Thomas Edison once said is 1% inspiration and 99% perspiration. The perspiration arises in preparing your mind—building the foundations upon which the building blocks of creativity are set. The historical accounts of truly great discoveries tend to focus on the final leap of intuition which made the breakthrough. Little mention is made of the months or years of researching the problem, wrestling with existing solutions and exploring all the ideas which eventually proved useless. In many of these cases it was the articulation of incorrect solutions which led, step by painful step, to the breakthrough. So it would seem to be with new product development. Creativity must be given the freedom to come up with unsuccessful ideas in order to discover the successful ones. Indeed, the number and quality of the ideas which are rejected is probably the best measure of a person's idea-generating capabilities. Taken to its logical extreme, the closer you get to thinking up every possible solution to a problem, the closer you will be to finding the best possible solution. Freedom to express creative ideas, including those which ultimately prove worthless, is, therefore, a virtue in new product development.

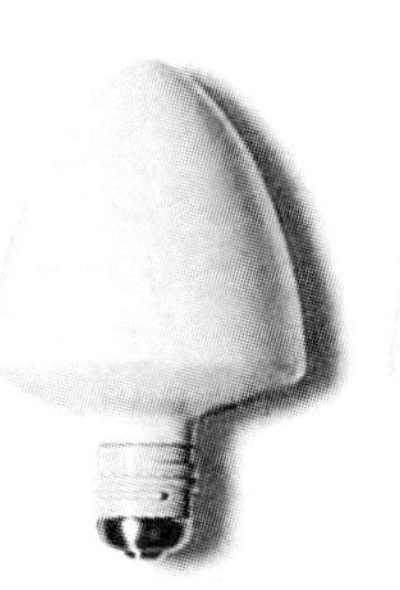
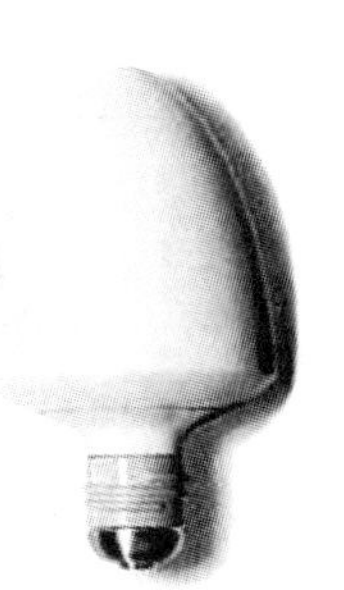

创造性表达的自由是设计的核心所在
Freedom of creative expression is at the heart of design.

拓展阅读

Extend and reading

Enzo Mari

Product+Furniture Designer(1932–)

One of the most thoughtful and intellectually，provocative Italian designers of the late 20th century. Enzo Mari has proved as influential to younger generations of designers as to his peers as a writer，teacher，artist and designer of products，furniture and puzzle games.

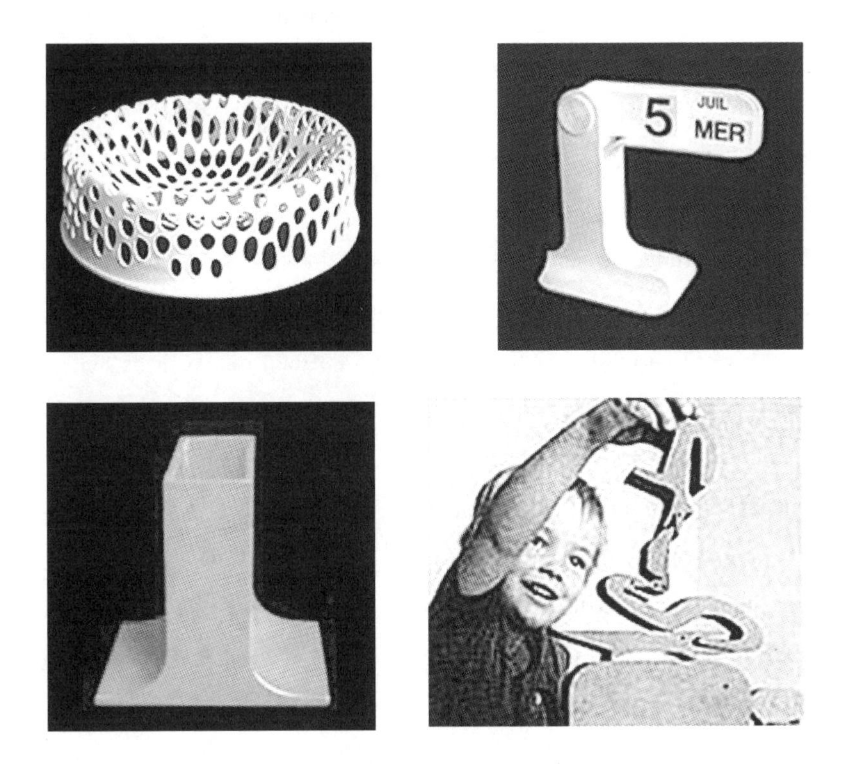

Born in Novara，Italy，in 1932，he studied classics and literature at the Academia di Brera in Milan from 1952 to 1956. As a student，Mari supported himself by working as a visual artist and freelance researcher. In a period when Italian design was flourishing as enlightened industrialists collaborated closely with designers to rebuild their businesses，he also became interested in design and painstakingly taught himself about it.

Mari's approach to design was predominantly theoretical. He was more concerned with its role in contemporary culture and relationship with the user than with becoming a design practitioner. After graduating in 1956 he opened a studio in Milan to continue his studies of the psychology of vision，systems of perception and design methodologies. These studies took physical form when Mari created three-dimensional models of linear elements and planes. Forced to earn a living, Mari made contact with the Italian plastic products manufacturer Danese and agreed to develop a series of mass-manufactured products.

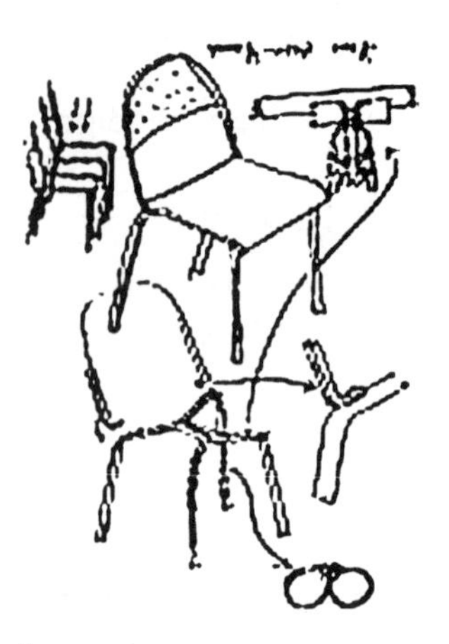

恩佐的草图
Enzo's sketch.

His first project for Danese was 16 Animali, or 16 Animals, launched in 1957. It was a wooden puzzle to which Mari applied his theories of problem-solving to create a group of simply carved animal shapes—including a hippo, snake, giraffe and camel—that join together to form a rectangle. The puzzle marked the start of a long collaboration between Mari and Danese, which continued at the turn of the 1960s with the development of containers and vases. Mari was determined to develop these products for mass production without compromising his belief that the outcome of each design project should be beautiful to look at and feel, while performing its function efficiently. Describing his philosophy as one of "rational design", he defined his work as being "elaborated or constructed in a way that corresponds entirely to the purpose or function".

Mari continued his experimental work in other areas of the visual arts notably by founding the Nuova Tendenza group of artists in Milan in 1963. Yet he was equally productive as a designer. In 1962, he began to work with Danese's signature material—plastic—in a six year-long project to develop a hat stand, umbrella stand and waste bin. By the end of the 1960s, Mari could manipulate plastic so skilfully that, in his hands, it attained a sculptural fluidity. One of his most accomplished plastic products was vase Model 3087 which Danese put into production in 1969. It was a reversible vase with a central cone to ensure that it functioned equally efficiently as a vase whether standing on its top or bottom. Made from glossy ABS plastic, Model 3087 had such a sensual shape that, to many design critics, it played a decisive role in persuading the public that plastic products need not necessarily be cheap and tacky.

While continuing his work as a product designer, he also turned to furniture. In 1971, Mari unveiled the Sof Sof chair for Driade in which a single removable cushion upholstered a simple welded rod frame. Equally ingenious was the 1975～76 Box for Castelli, a self-assembly chair consisting of an injection-moulded polypropylene seat and collapsible tubular metal frame which came apart to fit into a box, just as his puzzles interlocked into a rectangle.

By the early 2000s, as Mari entered his seventies, he won new commissions from Muji, the Japanese home store with a rationalist aesthetic remarkably empathetic to his own, and Gebrüder Thonet in Vienna, which invited him to create a contemporary version of its famous late 19th century bentwood chairs.

第10课 | 飞利浦设计公司的三个案例
Lesson 10 | Philips Design's three cases

词汇和短语
Words and phrases

studio (设计)工作室

humanize 赋予人性，人性化

multicultural 多元文化的

aspect 面貌，(问题的)方面

case 案例

brand 商标，品牌

audience 观众，听众

stigma 污名，耻辱

associated with 与……有关联的

mountain bike 山地自行车

LCD screen 液晶屏

head-up view 高阔的视野

battery charge 电池充电

option 选项，选购件

paradigm 范例

diagnostic equipment 诊断设备

clinical 临床的，病房用的

icon 图标，肖像

font 字体，字形

evolutionary 进化的，渐进的

wearable 耐久的

garment 衣服

sensorial 感知的，感觉的

multimedia 多媒体

enthusiasm 狂热，热心，积极性

overwhelming 压倒性的，无法抵御的

词汇联想与记忆

Association and memory of words

human	*n.*	人，人类
	adj.	人的，人性的，有同情心的
cultural	*adj.*	文化的
view	*n.*	景色，风景；观点，见解；观察，观看；意见，认为
	vt.	观察，观看
brand	*n.*	商标，牌子；烙印
	vt.	打火印；侮辱
equipment	*n.*	装备，设备，器材，装置；铁道车辆；(一企业除房地产以外的)固定资产；才能
option	*n.*	选项；选择权
media	*n.*	媒体

关键词汇和概念

Key words and concepts

飞利浦设计公司是世界上最大的、国际认可的设计工作室之一；它通过提供技术人性化的解决方案而创造价值。它的“未来设计”使其载誉世界。

Philips Design is one of the largest, international recognized design studios in the world, it creates value by providing solutions that humanize technology. Its “future design” makes it famed the world over.

课文

Text

飞利浦设计公司的三个案例

Philips Design's three cases

飞利浦设计公司是世界上最大的、国际认可的设计工作室之一，一支超过450名专业人士的创造性团队，通过提供人性化技术的解决方案而致力于创造价值。这是个由30个国籍人员组成的多元文化设计团队，很年轻——平均年龄34岁，而且40%是女性。其设计服务范围覆盖了商业创新的各个方面。

Philips Design is one of the largest, international recognized design studios in the world, a creative force of over 450 professionals dedicated to creating value by providing solutions that humanize technology. Its design team is a multicultural community representing 30 nationalities, young—average age is 34—and with 40% women. Its full range of design services covers all aspects of business creation.

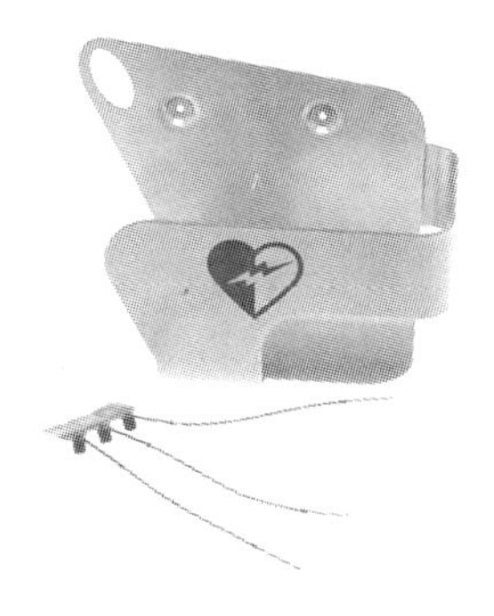

三个项目案例

Three project cases

a) 福特的“思索”自行车

a) Ford Th!nk Bike

客户的挑战：“思索”作为其新的环境友好品牌的一部分，福特想要开发一款电动自行车，通过不同凡响的设计应能让人认识到它与用户友好，能吸引广大观众，并能避免常与电动车辆联系在一起的坏名声。

Client challenge: as part of its new environmentally friendly brand Th!nk, Ford wanted to develop an electric bike that, through outstanding design, would be perceived as user-friendly and appealing by a wider audience and would avoid the stigma often associated with electric vehicles.

飞利浦设计的解决方案：“思索”自行车的造型应有便利感和山地车的外形，在舒适和运动体验之间达到平衡。这个定位给乘骑者以交通上好的控制性和高阔的视野。有个液晶屏显示诸如电池充电、速度表等信息，还有输入防盗代码可使自行车无法骑行的选购功能。

Philips Design' s solution：Th!nk Bike is styled for convenience and mountain bikes with a balance between comfort and sporting experience. The position provides the rider with good control and a head-up view in traffic. An LCD screen displays information such as battery charge，speedometer，as well as the option to disable the bike by entering an anti-theft code.

结果：“思索”自行车在2000年底特律汽车展上首次展出。它帮助福特公司从自行车开始导入电动车辆市场，创造了一个新的范例。

The result：the Th!nk Bike was first presented at the Detroit Motor Show 2000. It helped to create a new paradigm for Ford to introduce itself in the market of electric vehicles，starting with bicycles.

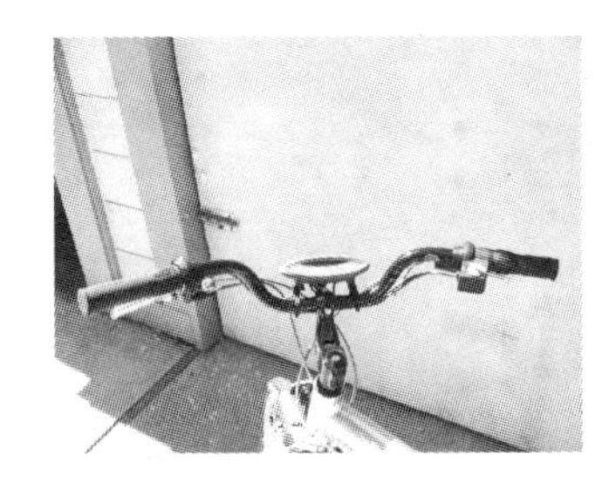

“思索”自行车
Think Bike

b) 飞利浦医疗系统的形象识别设计

b) Philips Medical System (PMS) Identit

客户的挑战：飞利浦医疗系统为卫生保健机构提供诊断设备和服务。其主要服务对象是临床用户、购买过程(医院管理者)中的决策者和患者。产品的特点最为重要的是品牌与用户之间的完全信任。

Client challenge：PMS provides diagnostic equipment and related services to the healthcare community. Its main service objects are clinical users，the decision makers in a purchasing process (hospital management)and the patients. The nature of the products makes it of the utmost importance that complete trust exists between the brand and the users.

飞利浦设计的解决方案：飞利浦医疗系统在传统上依赖于飞利浦设计公司通过一致的产品形象识别来创造和维持其与消费者进行的对话。协调人员在公司层面上从形态语言、色彩、图标、符号、产品的图形、产品的品牌、字体、控制和显示等方面进行了规划开发。这些规划经过正规的评估。

Philips Design’s solution：PMS has traditionally relied upon Philips Design to create and maintain the ongoing dialogue with its customers through a consistent product identity. Harmonisation programers are developed at corporate level for form language，colours，icons，symbols，product graphics，product branding，fonts，controls and displays. These programmes are evaluated on a regular basis.

结果：在最近五年飞利浦医疗系统得到了35个奖项，在产品设计和用户界面设计上被认为是顶级的。1999～2000年飞利浦医疗系统产品形象识别方面进行的渐进步伐加强了这一地位。

The result：over the last five years，PMS has received 35 awards and is considered best-in-class for product design and user interface design. An evolutionary step in the PMS product identity in 1999～2000 reinforced this position.

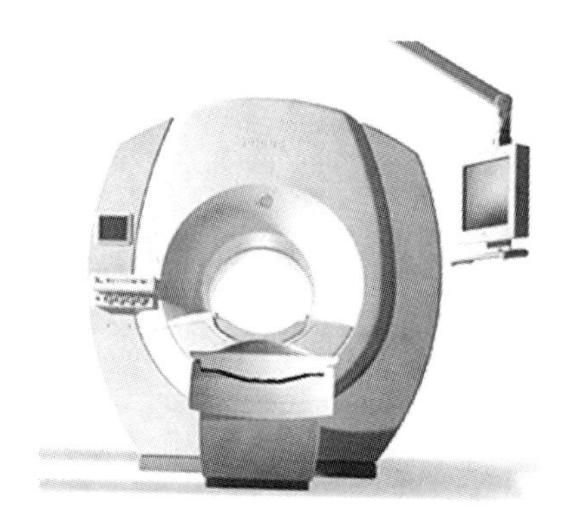

拓展阅读

Extend and reading

The project case c：

Levi's ICD+Launch

Client challenge：ICD+was the first wearable electronics garment to be put on the market for consumers. Co-design by Philips Design and Levi's, a famous dress company, developed by Philips Design and Philips Research, the garment was launched in 2000.

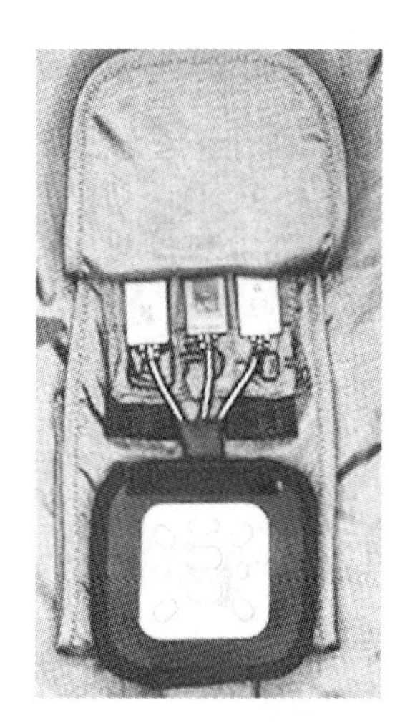

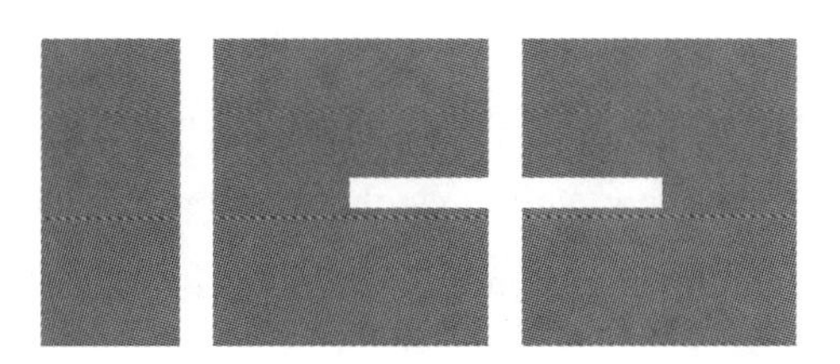

The challenge for Philips Design, in charge of the launch event, was to convey the idea that the garments were not only innovative, but also the start of a whole new lifestyle.

Philips Design' s solution：the launch event programme allowed the audience to experience the new lifestyle as well as the new product.

The sensorial multimedia show suggested comfort and freedom through images and sounds. The lectures on the history of the merger of technology and fashion addressed the mind, whilst the body was involved when editors had the chance to try the garments on.

The result：the enthusiasm for the product was overwhelming, and the launch event had a snowball effect in terms of creating press attention：after only a few months, hundreds of articles had been written whilst ICD+ was already heading towards its second generation.

About the Philips Design

We are a global community committed to enriching the design process and delivering competitive value to our clients. We create design solutions focused on personal growth, so people can exist in harmony with each other and with their environment.

There are several hundred of us, scattered across 12 sites worldwide. This gives us a fascinating insight into emerging local trends and developments, some of which can have significant consequences far beyond their region. It also makes us one of the largest design studios in the world.

We embrace technology, but as a means to an end. It is an enabler, a way of achieving a better quality of life. Our visionary approach—enriching design with human sciences, user research and always with a clear people focus—allows us to shape technology in a way that answers people's existing and latent needs.

It is our intention to design solutions that harness technology so as to genuinely improve the quality of people's lives and make them happier. We constantly stress the need for relevance, context and genuine appeal, based on expressed user preferences, in everything we do. Is it possible to create a "landscape of happy objects"? Sure.

But is technology surely becoming too intrusive? Only if we let it. Since embarking on Ambient Intelligence projects in the early 1990s, we have always tried to strike a balance so people are empowered—and not overpowered—by technological innovation.

And, naturally, we believe our design solutions should be responsible and sustainable. Whether from a business, environmental, personal, social or moral point of view, there isn't really any other feasible option.

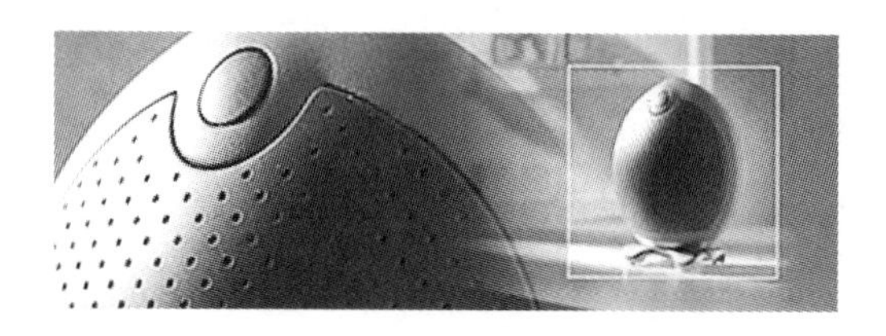

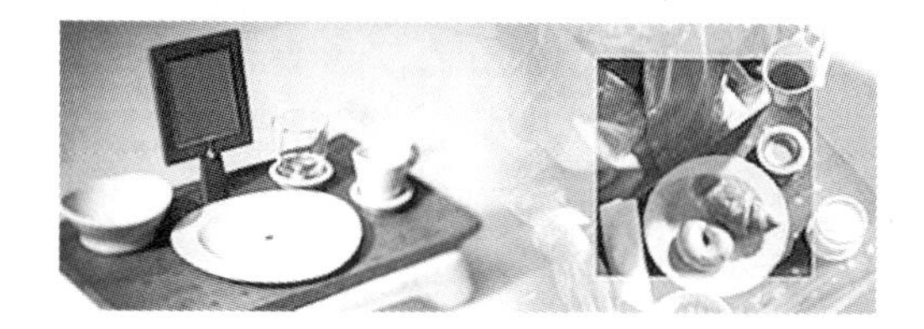

第11课 | 青蛙设计公司
Lesson 11 | Frog Design Company

词汇和短语
Words and phrases

motivation　动机

emotion　情绪，情感

commission　委托；任命

hit　击中；(演出等的)成功

strike　(过去时struck)拍板

entice　诱惑，诱使

synonymous(with...)　与……同义的

pitch in　努力投入工作

switch (to)　转换，转变(到……)

there's no point　毫无意义

flatware　(刀、叉、匙等扁平)餐具

spoil　扰乱，搞糟

no matter how　无论怎样

appeal　要求，诉求，呼吁

laud　赞誉，赞美

be lauded as...　被赞誉为……

sculpture　雕塑

time and again　屡次，反复不断地

revenue　收入，税收

soar　剧增，高飞

词汇联想与记忆

Association and memory of words

motive　*n.*　动机，目的

　　　　adj.　发动的，运动

strike　*n.*　罢工，打击，殴打

　　　　vt.　打，冲击，罢工；打动

switch　*n.*　开关，电闸，转换

　　　　vt.　转换，转变

point　*n.*　点，尖端，分数，要点

　　　　vt.　弄尖；指向；指出；瞄准，加标

　　　　vi.　指，指向，表明

ware　*n.*　陶器，器皿

matter　*n.*　事件，问题，物质，内容，实质，原因，文件

　　　　vi.　有关系，要紧

关键词汇和概念

Key words and concepts

源于德国的青蛙设计公司现在是世界上最大的设计公司之一。青蛙公司的设计理念是：好的设计源自于文化。“形式服从情感”。

Originally from Germany，Frog Design is one of the largest design companies in the world now. Frog Design's philosophy on design：good design is culturally relevant. “Form Follows Emotion.”

课文

Text

青蛙设计公司

Frog Design Company

在青蛙设计公司，好的设计来自于我们的技能、经验和直觉以及对使用者需求和动机的深刻理解的结合……这就是为什么我们说“形式服从情感。”青蛙的设计理念是：好的设计源自

于文化。

At Frog Design, good design comes from combining our skill, experience, and intuition with a deep understanding of users' needs and motivations...which is why we say, "Form Follows Emotion." Frog Design's philosophy on design: good design is culturally relevant.

青蛙设计公司是1969年由哈特姆特·埃斯陵格与合伙人安德列斯·哈格及乔格·斯泼赖在德国创建的。那时埃斯陵格说过:“我想让人类微笑”。

Frog Design was founded by Hartmut Esslinger, along with partners Andreas Haug and Georg Spreng, in Germany in 1969. At that time Esslinger said "I wanted to make people smile."

埃斯陵格职业的早期突破始于德国电子巨商威格公司的一次委托。这是一次巨大的成功。这是他为该公司设计的100多个产品的第一个。

Early in his career, Esslinger's break came with a commission from the German electronics giant Wega. It was a big hit. It was the first of more than 100 products he designed for the company.

几年后，威格公司由索尼收购。这次，他的第一个大成功是黑盒子的索尼特丽珑电视机。下一个突破性的项目是在斯蒂夫·乔布斯，苹果的创始人之一，在搜寻设计公司时来到的。一笔几百万美元的交易拍板了，这诱使埃斯陵格在加利福尼亚州开设了办事处。青蛙设计公司逐渐成为创新、想象力和成功的代名词。

A few years later Wega was bought by Sony. This time his first success was the black-box Sony Trinitron TV. The next ground-breaking project came along when Steve Jobs, co-founder of Apple, was searching for a design company. A multimillion-dollar deal was struck, which enticed Esslinger to establish a California office. Gradually Frog Design became synonymous with innovation, vision, and success.

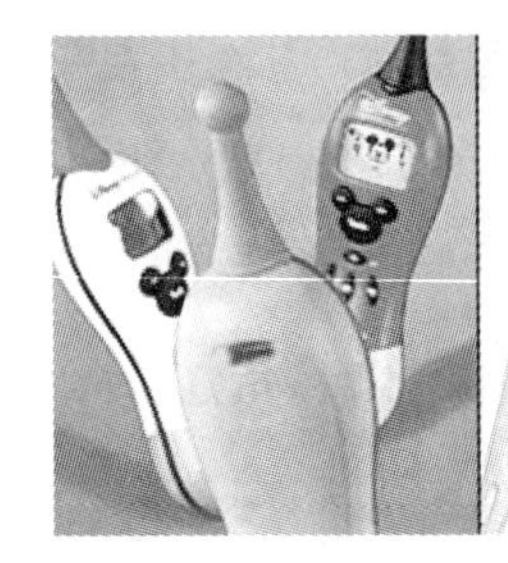
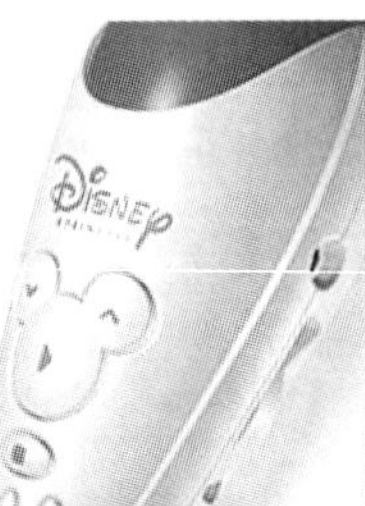

青蛙公司设计的产品
Products designed by Frog Desgh

青蛙创造了整合的消费者体验。埃斯陵格总是鼓励每个人在项目上作出贡献——工程师在设计上有发言权，设计师则要在材料和系统上努力投入。荣誉不给任何单个个人，每个项目都是团队工作的产物。

Frog Design creates integrated customer experiences. Esslinger has always encouraged the contributions of everyone working on a project—engineers have their say on design, and designers pitch in on materials and systems. No single person is credited; each project is the product of team work.

而且在许多综合项目里，例如像视窗XP项目，通过品牌、产品和数字的整合，工业设计师与图形和数字设计师一起在三维空间里工作。开始时他们混合起来工作，然后在项目进展中每个人可以转回到他们的专门领域里。

Furthermore, with the integration of brand, product and digital, industrial designers work with graphics and digital designers within the 3-D space in many of their integrated projects, such as the Windows XP project. The mix happens at the start, then as the project progresses individuals can switch back to their areas of expertise.

对于一个从概念阶段直至销售的项目而言，它可能包括产品设计、工程、生产、制图、标志、包装和数字媒体，“如果产品不工作，那么在图形设计上花费钱就毫无意义”。在德国汉莎航空公司的项目里，青蛙就注视其整体运行状况，从售票到抵达目的地。结果每件事都予以重新设计——从机场标志到飞机内饰、到餐具。一个元素错了，就可能扰乱整个系统。

For a project, from the idea stage right through to the sale, that might include product design, engineering, production, graphics, logos, packaging, and digital media, "there's no point in spending a fortune on graphic design if the product doesn't work." With Lufthansa, Frog Design looked at their entire operation from ticket sales to destination. In the end, everything was redesigned—from the airport signage to the plane interiors, to the flatware. When one element is wrong, it can spoil the whole.

埃斯陵格相信设计应该永远包括某些额外的东西："魔力在于制造商和消费者得到了某些他们没有预期到的好东西之时。"对于埃斯陵格而言，一个设计无论怎样雅致和具有好功能，如果不能够在更深的层次上诉求我们的情感，就不会在我们生活里赢得地位。

Esslinger believes design should always include something extra: "The magic is when both the manufacturer and consumer get something good that they don't expect." For Esslinger, no matter how elegant and functional a design, it will not win a place in our lives unless it can appeal at a deeper level, to our emotions.

青蛙的许多设计在它们投放市场几个月内就达到了杰出的状态。索尼的特丽珑重新诠释了电视机的设计；奥林帕斯的BX-40显微镜一直被誉为工业的雕塑；华美的汉斯格鲁赫淋浴喷头安装在欧洲所有的浴室里，至今已售出了1500万个以上。

Many of Frog Design's designs have reached classic status within months of their launch. The Sony Trinitron redefined television design; the Olympus BX-40 microscope has been lauded as industrial sculpture; the colorful Hansgrohe showerhead is found in bathrooms all over Europe and has sold more than 15 million units to date.

青蛙设计已经屡次帮助一个公司重新发现它的情感潜力。在苹果公司想要将麦金托什计算机投放市场时，它让青蛙来完成设计；使年收入从1982年的7亿美元飚升到1986年的40亿美元。同样，青蛙公司在标志、指示装置以及包装方面所做的工作使罗技公司的年收入急剧攀升，从1988年的4300万美元升到1995年的2亿美元，保证了该公司在市场上占领首位。

Time and again Frog Design has helped a company rediscover its emotional potential. When Apple wanted to launch the Macintosh computer it called in Frog Design to complete the design; revenue soared from $700 million in 1982 to $4 billion in 1986. Likewise work on the corporate logo, pointing devices, and packaging for Logitech saw revenue climb very steeply from $43 million in 1988 to over $200 million in 1995, securing the company its number-one market position.

今天，青蛙设计继续是设计方面的领先者，为准备变革的公司寻求下一个突破项目。

Today, Frog Design continues to be a leader in design and is looking for that next ground-breaking project in a company ready for change.

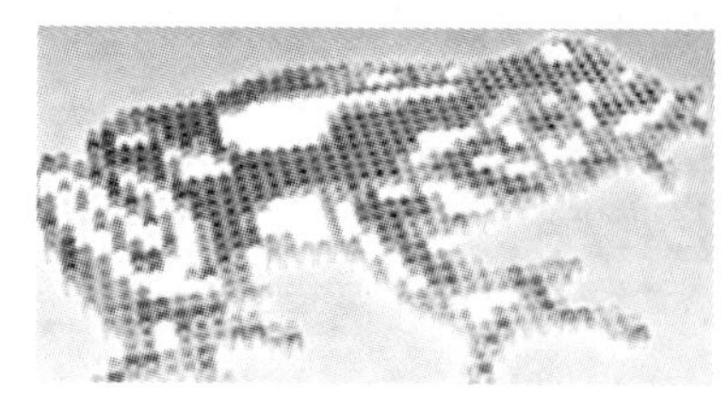

青蛙公司的标志
The siga of Frog Design

青蛙公司的客户企业
Clients of Frog Design

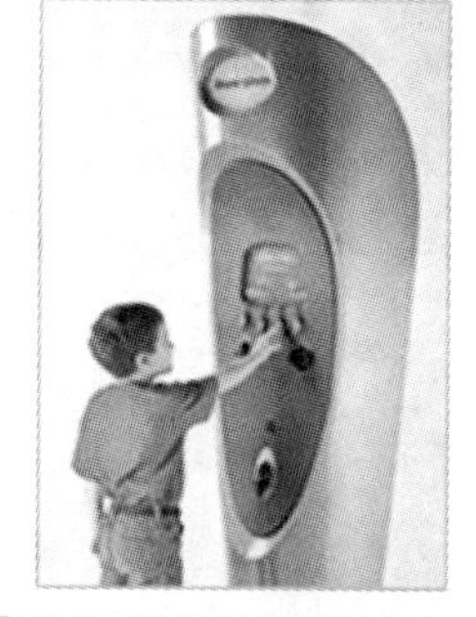
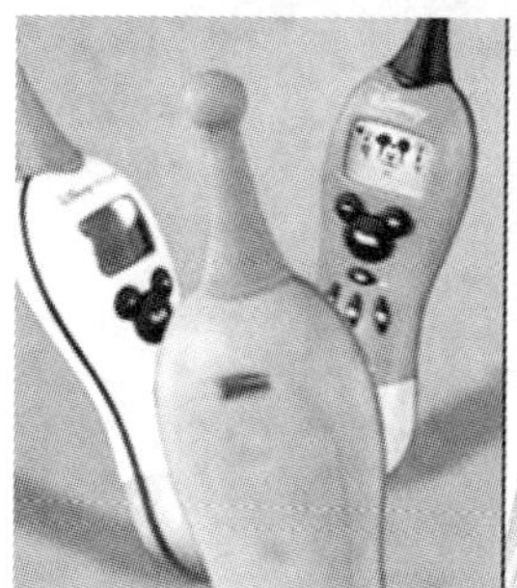

青蛙设计公司为迪斯尼乐园设计的高速艇、多媒体亭和电子玩具
The cruise boat, multimedia kiosk and electrical toys

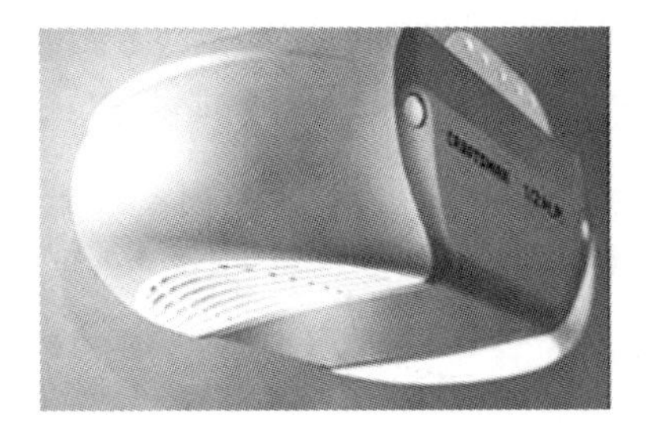
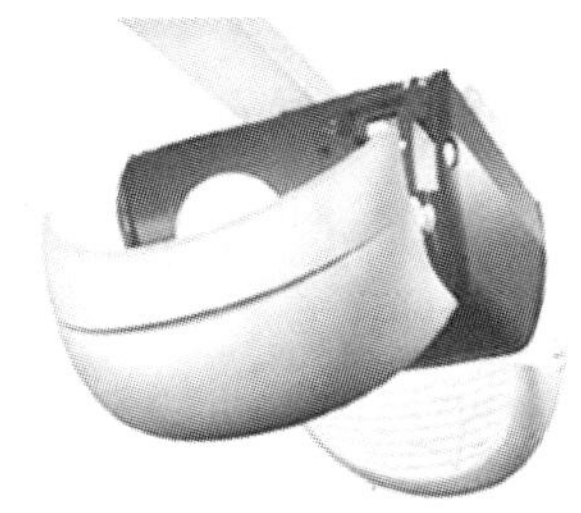
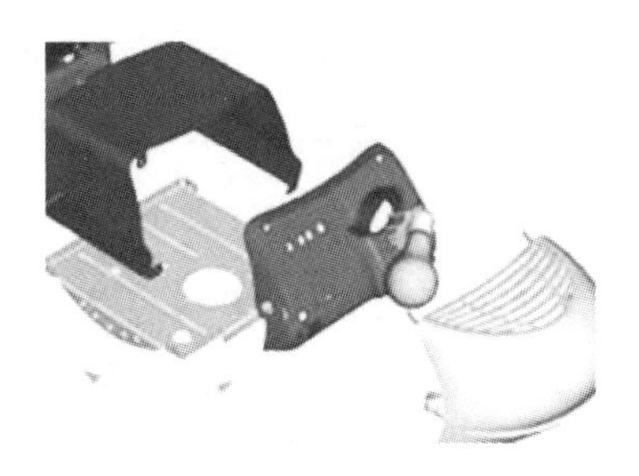

青蛙设计的升降单元的渲染图和爆炸图
3-D effect and exploded drawings for a lift unit designed by Frog Design

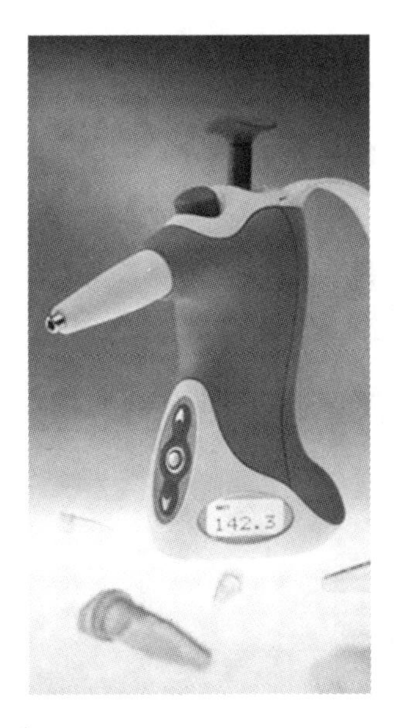

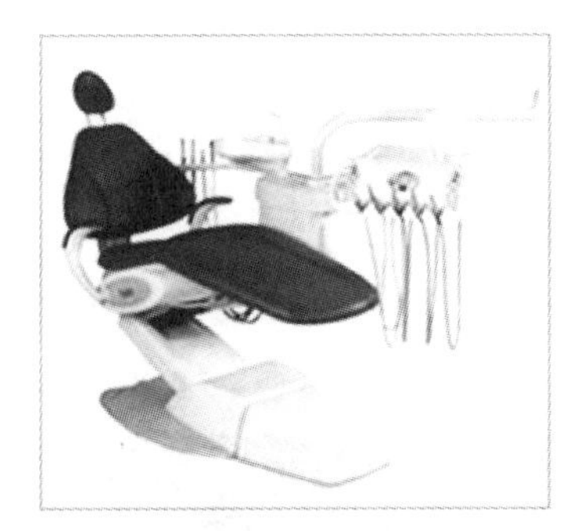
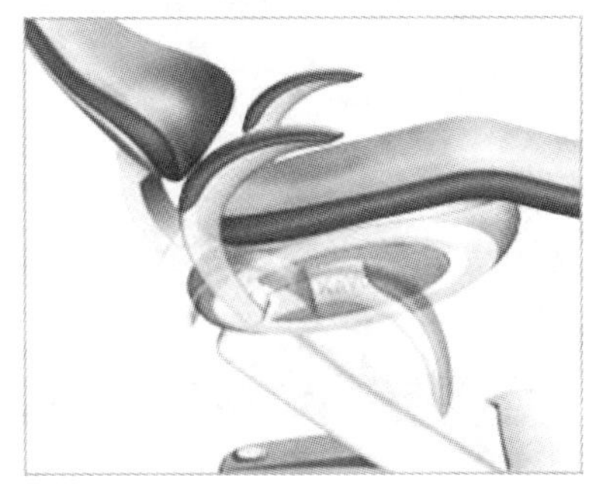

青蛙为客户设计的桌上电子产品、彩电和牙科zxx手术床
Electrical appliances and dental chair

拓展阅读

Extend and reading

Frog and flextronics

Who is flextronics?

Flextronics is the leading Electronics Manufacturing Services(EMS)provider focused on delivering operational services to technology companies. With fiscal year 2004 revenues of USD$14.5 billion, Flextronics is a major global operating company with engineering, manufacturing, and logistics operations in 32 countries and five continents. This global presence allows for manufacturing excellence through a network of facilities situated in key markets and geographies that provide customers with the resources, technology, and capacity to optimize their operations. Flextronitcs' ability to provide end-to-end operational services that include innovative product design, test solutions, manufacturing, IT expertise, network services, and logistics has established the Company as the leading EMS provider.

"...We're jumping ahead of the curve again."

—— Hartmut Esslinger

Why Frog and Flextronics?

The Frog+Flextronics will create a unique soup-to-nuts product-development outsourcer that will provide a look and user experience that a customer can fall in love with—together, we'll also get those products to market more efficiently, saving time and money.

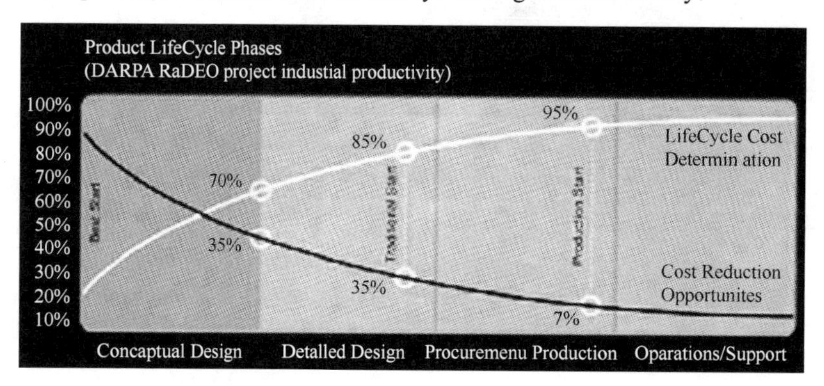

"Come Together"

Hartmut Esslinger, Co-CEO of Frog Design and Michael Marks, CEO of Flextronics share a vision:that today's and tomorrow's products will have to be more profitable to all players from retail to brand owners to distribution to manufacturing by providing superior customer experiences. Together, Frog and Flextronics can now integrate the entire PLM-cycle from "front-to-end" without compromising their core competencies: world-leading design and innovation combined with world-leading technology, production and logistics.

Frog understands innovation

For over 30 years, Frog has repeatedly distinguished itself by producing market-leading products, services and brand enhancements. We create innovative solutions for companies competing in rapidly changing marketplaces.

Frog understands business

Our unique process of concurrent, multidisciplinary innovation is specifically tailored to the business needs of our clients. We understand the challenges businesses face, and our innovation process allows us to achieve unprecedented ROI and time to market. We create business impact through physical, digital and emotional innovation.

Physical Innovation

We help our clients dramatically increase ROI and brand equity through innovative, timely product releases. Our work in product design for Apple(Macintosh), Sony(Trinitron)and Ford(Th!nk Vehicle)is legendary, and the business impact unmistakable.

Digital Innovation

Today' s consumer demands a strong digital presence. Our work for Dell. com, Microsoft Windows XP and i2 demonstrates our expertise in user interface. We can put it to work for you, increasing your profits through effective execution of e-commerce strategy.

Emotional Innovation

Our branding team can help your firm increase customer loyalty through our proven strategic branding process. We did it for Acura, Oracle and SAP, and we can do it for you.

第12课 | 苹果公司的设计师乔纳森·艾夫
Lesson 12 | Apple's Designer Jonathan Ive

词汇和短语
Words and phrases

consultancy　顾问，咨询(工作)

drab　土褐色的，单调的

opaque　不透明的

beige　米色，浅褐色的

close-knit　组织严密的

passionate　充满热情的

laptop　膝上型手提电脑

slim　苗条的，纤细的，小巧的

set-up　组织机构，结构

humbling　令人羞辱的

context　上下文关系，文脉；背景

commit　效忠，献身(to...)

more than ever　益发，愈加

(be) reliant on　依赖于，依靠于

massively　严重地，结实地

loyalty　忠诚，忠心

heritage　遗产，传统

consequent　作为结果的，随之发生的

imperative　势在必行的，绝对必要的，强制的

polymer　聚合物

shoot(ing)　射击；注射，注塑

moulding　浇注，浇铸，模铸

twin-shot　双塑注射，双料注塑

fastener　纽扣，按钮，卡扣

dense　密集的；极度的

adhesive　黏合剂

laser　激光

weld(ing)　焊接；焊缝

词汇联想与记忆

Association and memory of words

passion　*n.*　激情，热情

reliance　*n.*　信心，依靠，依靠的人或物

关键词汇和概念

Key words and concepts

苹果电脑公司刮起的“透明风”始于计算机iMac，其后G4、G5、iPod等，接连不断。这些都是设计师乔纳森·艾夫的杰作。

“Clarity” prevalence blown by Apple Computer, started from Computer iMac, being in succession G4, G5, iPod and so on. These are all masterpieces of designer Jonathan Ive.

课文

Text

苹果公司的设计师乔纳森·艾夫

Apple's Designer Jonathan Ive

第一部分

Part I

乔纳森·艾夫1967年生于伦敦，在纽卡斯尔理工大学(今诺森比亚大学)学习艺术设计；后合伙创办了丹吉尔设计咨询公司。 1992年，他的一个客户——苹果公司——在美国加利福

尼亚州的总部给他提供了一个职位。在与苹果公司创始人之一斯蒂夫·乔布的亲密合作期间，艾夫开发了iMac。在此之前，单调的计算机世界里的新产品都是用不透明的灰色或米色塑料外壳所装裹；通过引入色彩和灯光，iMac改变了产品设计，第一年就售出了200多万台。

Born in London in 1967, Jonathan Ive studied art and design at Newcastle Polytechnic(now Northumbria University)before co-founding Tangerine, a design consultancy. In 1992, one of his clients-Apple-offered him a job at its headquarters in California, U.S.Working closely with Apple’s co-founder, Steve Jobs, Ive developed the iMac. As well as selling more than 2 millions units in its first year, the iMac transformed product design by introducing colour and light to the drab world of computing where, until its arrival, new products were encased in opaque grey or beige plastic.

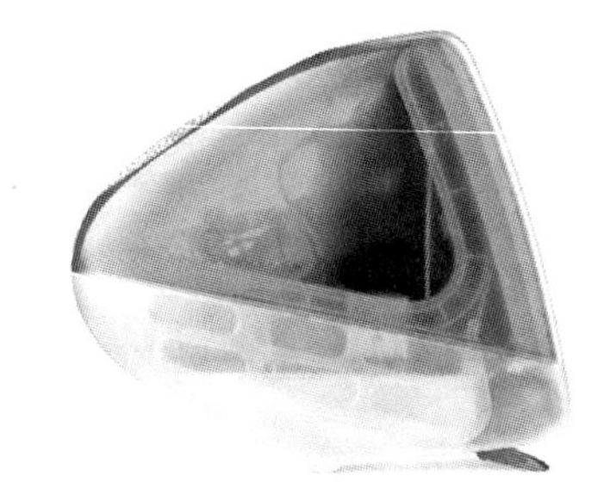

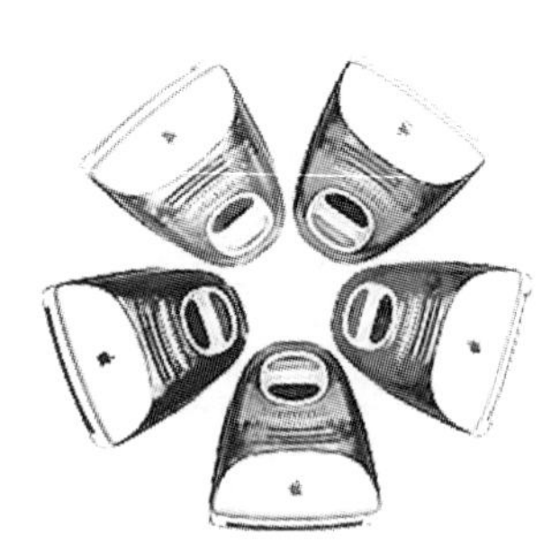

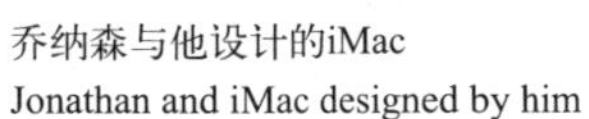

乔纳森与他设计的iMac
Jonathan and iMac designed by him

艾夫和他在苹果公司里组织严密的设计师团队此后使用了同样的横向思维方法以及对细节充满热情的关注来开发同样创新的新产品，像Cube、iPod和世界上最轻巧的17英寸膝上型电脑PowerBook G4，以及超小的iMac G5。

艾夫设计的iPod和iSub
The iPod and iSub designed by Ive

Ive and his close-knit team of designers at Apple have since applied the same lateral thinking and passionate attention to detail to the development of equally innovative new products such as the Cube, the iPod and the PowerBook G4, the world' s lightest and slimmest 17 inch laptop, and the ultra-slim iMac G5.

第二部分
Part II
问与答
Q&A

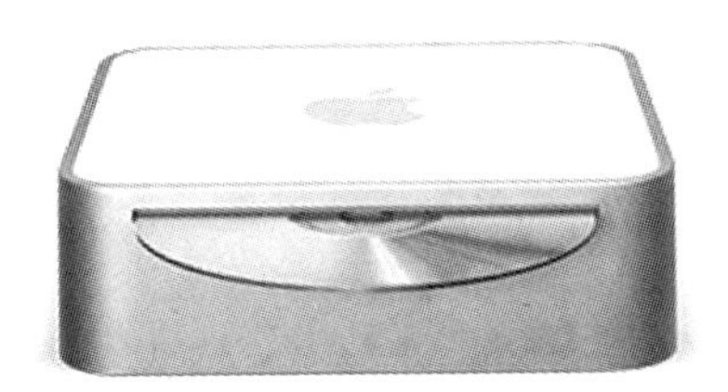

问：为一个公司设计的优点是什么？那缺点呢？苹果公司组织机构的特征是什么？

Q：What are the advantages of designing for one company? And the disadvantages? What are the particular characteristics of the set-up at Apple?

答：关键是，不仅一个公司的领导层要清晰地理解其产品及设计的任务，而且开发、市场和营销队伍也要同样忠诚于同一个目标。我益发明白的是，我们用设计所要完成的事情大量地依赖于许多不同的团队为解决同一个问题的投入，依赖于他们为同样的目标工作。我喜欢成为比设计更大的某件事情的一部分。我忠诚于苹果公司，也相信苹果公司除了设计之外还有其他冲击力，这是重要的。我还觉得因活着而负有的责任，有时候因所做事情的结果而感到相当痛苦。

A：Not only is it critical that the leadership of a company clearly understands its products and the role of design, but that the development, marketing and sales teams are also equally committed to the same goals. More than ever I am aware that what we have achieved with design is massively reliant on the commitment of lots of different teams to solve the same problems and on their sharing the same goals. I like being part of something that is bigger than design. There is a loyalty that I have for Apple and a belief that this company has an impact beyond design which feels important. I also have a sense of being accountable as we really live, sometimes pretty painfully with the consequences of what we do.

问：苹果定义的产品设计质量是什么？这与你到那儿之前苹果的设计传统关联到什么程度？

Q：What are the defining qualities of the design of an Apple product? To what degree are they related to the design heritage of Apple before your arrival there?

答：在20世纪70年代，苹果谈论的是技术和艺术的交叉。我认为产品的质量实际只是公司创立时所确立的远大目标的一个结果。这样定义的质量涉及使用：容易和简便。除了关注绝对必要的功能之外，我们还认为，产品在传统的功能观念之外还具有其他重要意义。

A：In the 1970s，Apple talked about being at the intersection of technology and the arts. I think that the product qualities are really consequent to the bigger goals that were established when the company was founded. The defining qualities are about use：ease and simplicity. Caring beyond the functional imperative，we also acknowledge that products have a significance way beyond traditional views of function.

问：你说过，历史地看，设计的进步一直由新材料的开发所驱动。现在什么新材料最令你兴奋？

Q：You have said that，historically，advances in design have been driven by the development of new materials. Which new materials excite you most now?

答：材料、工艺、产品机构和结构是设计的巨大驱动力。聚合物的进步意味着我们现在可以创造复合物以满足非常特殊的功能目标和要求。从过程的观点看，我们现在已经能够用塑料做出以前被告知是不可能做到的事。双塑注射材料——两种不同的塑料一起注入模具或者塑料—金属混注，真的给了我们许多以前不会存在的功能和形态机会。iPod 是用双塑注射制造的，没有扣件，也没有电池盖，使我们能够创造一种极度密封的设计。金属成型，特别是用先进的粘结剂以及用激光焊接来连接金属的新办法，也是现在令人兴奋的领域。

A：Materials，processes，product architecture and construction are huge drivers in design. Polymer advances mean that we can now create composites to meet very specific functional goals and requirements. From a processing point of view we can now do things with plastic that we were previously told were impossible. Twin shooting materials—moulding different plastics together or co-moulding plastic to metal gives us a range of functional and formal opportunities that really didn’t exist before. The iPod is made from twin-shot plastic with no fasteners and no battery doors enabling us to create a design which was dense completely sealed. Metal forming and，in particular，new methods of joining metals with advanced adhesives and laser welding is another exciting area at right now.

拓展阅读

Extend and reading

Jonathan　Ive

Given the buzz attached to his name in the hallowed halls of Apple, Jonathan Ive might be expected to be something of an egomaniac. This shaved-headed, soft-spoken Brit is anything but. The only time you'll hear him use the word "I" is when he's naming some of the products he helped make famous: iMac, iBook, iPod.

Yet for all Ive's attempts to give away the credit to a design team he assembled, his fingerprints are all over Apple's five-year-long radical shift in hardware design. When the Cupertino, Calif., computer maker hired Ive in 1992, it was still cranking out beige-box desktops and creaky black plastic PowerBooks. When Steve Jobs appointed Ive vice president of industrial design in 1997, everything changed.

Ive began using materials, shapes and colors that had never been seen in the industry before. The original iMac broke the beige-box mold with curves, candy colors and a carrying handle. No one else has even tried to build a computer like the latest iMac, with a flat screen on a movable metal stalk. No one else has made inch-thin laptops out of titanium or aircraft-grade aluminum. Or a keyboard, now on the latest PowerBook, that knows when it gets dark and lights up accordingly.

It's not as if Ive were trying to be radical, he's just sweating the details. "You deal with the needs and problems the product has," he says. "The result is often something you didn't expect." Unlike some designers, Ive uses his own products after they're finished. Doing so gives him ideas for updates to later models. Tweaks like making the movable track wheel on the iPod sensitive only to touch, so it doesn't jog when you do, are all the more successful because even though they're incredibly complex, we barely notice them. You could say the same for the self-effacing Ive.

Biography

1967

Born in London, where he spends his childhood.

1985

Studies design and art at Newcastle Polytechnic(now Northumbria University).

1989

Becomes a partner at Tangerine, a London-based design consultancy where he works on a wide range of products from power tools to wash basins.

1992

Moves to San Francisco to join the Apple design team.

1998

Appointed vice president of industrial design at Apple.

Launch of the original iMac, which sells 2 million units in its first year.

1999

Introduction of the Apple iBook, the 22″ Cinema Display, PowerMac G4 Tower and iSub.

2000

Launch of the Apple G4 Cube.

2001

Apple introduces the Titanium PowerBook G4 and the iPod portable MP3 player.

2002

Launch of the new sunflower-inspired iMac with 15″ and 17″ floating screens. Introduction of the eMac, a version of the iMac specially developed for use in the education sector.

2003

Apple launches the 12″ PowerBook and the 17″ PowerBook, which at 1″ thick and 6.8 lbs is the world's slimmest and lightest 17″ notebook computer. Wins the Design Museum's first Designer of the Year Prize.

2004

Launch of the multi-coloured iPod mini and ultra-slim iMac G5.

2005

Appointed senior vice president of design at Apple. Launch of the Mac Mini.

"There's an applied style of being minimal and simple, and then there's real simplicity," he said. "This looks simple, because it really is."

第4单元
Unit 4

第13课 | 设计规范
Lesson 13 | Design specification

词汇和短语
Words and phrases

opportunity specification　机会规范，机会特性

design specification　设计规范

market research　市场调研

competing product　竞争产品

market need　市场需求

business opportunity　商业机会

break-even point　盈亏平衡点

technology auditing　技术审核

price position　价格定位

target cost　目标成本

financial model　财务模型，财务模式

quality control　质量控制

词汇联想与记忆
Association and memory of words

specification　*n.*　详述[常 pl.]，规格，说明书，规范

opportunity　*n.*　机会，时机

research　*n.*　研究，调查

vi.　研究，调查

financial *adj.* 财政的，金融的

position *n.* 位置，职位，形势，阵地

vt. 安置，决定……的位置

quality *n.* 质量，品质，性质

关键词汇和概念

Key words and concepts

设计规范是产品企划的关键工作之一，是产品目标的说明，它描述的是特定技术性能。设计规范是用于产品开发的关键性质量控制文件。

The design specification is one of key works of product planning，a description of product objectives，and describes the special technology properties. The design specification is the key quality control document for product development.

课文

Text

设计规范

Design specification

产品企划包括锁定机会、进行市场调研、分析竞争产品、提议新产品、拟定机会特性和设计规范等一系列工作。

Product planning includes spotting an opportunity，conducting market research，analyzing competing products，proposing a new product and drawing up both an opportunity specification and a design specification.

机会特性是一份简明扼要的书面文件，它描述了所提议的新产品的市场需求以及该产品所提供的商业机会。首先，这个文件会被管理者用来作为决定是否开始开发新产品的基础；接着，它又被用作一个概要文件让产品开发团队集中精力于他们要开拓的机会上。一个有效的机会特性包含了所提议新产品的以下两个方面：①叙述所提议新产品蕴含的商业机会，集中于产品核心利益的陈述。②根据计划销售情况、超过生产成本和开发费用后的边际利润以及计划的盈亏平衡点(在这点之上开发成本将可回收)，给出该商业机会的正当理由。

An opportunity specification is a concise written document describing the market need for a proposed new product and the business opportunity presented by that product. It is used，firstly，by management as a basis for deciding whether or not to make an initial commitment to the new product and subsequently, as a summary document to keep the product development team focused on the opportunity they are setting out to exploit. An effective opportunity specification covers two aspects of the proposed new product: ①It describes the business opportunity presented by the proposed new product, focusing on the product' s core benefit proposition. ②It provides a justification of that business opportunity in terms of projected sales, margin over manufacturing costs, development costs and projected break-even point(at which development costs will be recovered).

核心利益陈述是一份阐述消费者乐意购买该新产品而不是竞争产品的主要好处的简单明了的建议。因此，这份建议书的特别之处就在于竞争产品的分析，市场需求的调研及技术审核。

A core benefit proposition is a simple and concise proposition of the main benefit enjoyed by the customer from buying the new product instead of competing product. Therefore, this proposition is identified through the use of competing product analysis, market needs research and technology auditing.

参数 Parameter	维度 Dimension	竞争产品变量 Competing product variable	注解 Comments	理想产品变量 Ideal product variable
瓶子 Bottle	材料 Material	高密度聚乙烯 HDPE	再生循环性好 Good for recycling	高密度聚乙烯 HDPE
瓶子 Bottle	可循环材料的百分比 % from recycled material	0%	必须多使用可循环材料 Must use more recycling material	最少占40% 40% at least
瓶子 Bottle	质量(克) Mass(g)	105	能用较少材料 Could use less material	100克以下 Less than 100g
瓶盖 Cap	材料 Material	聚丙烯 Polypropelene	不同材料 Different materialt	高密度聚乙烯 HDPE
标签 Label	材料 Material	纸 Paper	不同材料 Different material	直接印到瓶子上? Direct printing on to bottle?
标签油墨 Label ink	颜料种数 No. of dyes	4	人造合成颜料 Synthetic dyes	2

核心利益建议是一份阐述消费者乐意购买该新产品而不是竞争产品的主要好处的简单明了的建议

A core benefit proposition is a simple and concise proposition of the main benefit enjoyed by the customer from buying the new product instead of competing product

以下是一些核心利益建议的实例：

Examples of core benefit proposition include：

● 美国运通的旅行支票——声誉好；到处都可使用，如有丢失可迅速得到更换和全面的保护。

● American Express Traveller's Cheques—prestigious; accepted everywhere，prompt replacement and complete protection if lost.

● 惠普激光打印机——在多种材料上无噪声地打印高质量文件；使用和维护简单、可靠灵活。

● Hewlett Packard Laserjet—quietly prints documents with excellent print quality on several media; easy to use and maintain，reliable and flexible.

● Silkience的自调整洗发水——对头发的不同部分给出恰当的清洁处理(自动地)。

● Silkience Self-Adjusting Shampoo—a shampoo which provides the appropriate amount of cleaning treatment(automatically)for different parts of your hair.

设计规范是对产品目标的说明，它描述的是特定的技术性能。设计规范是用于产品开发的关键的质量控制文件。它决定了产品必须做以及应该做的工作；它设定了一个标准，在质量控制的审查期间决定是否允许新产品继续开发或给取消。对设计团队来讲，设计规范也应该是一份实用而有益的指南，以保证在产品开发过程中产品设计不会有任何“遗漏”。

The design specification is a description of product objectives，and describes the special technology properties. The design specification is the key quality control document for product development. It determines what the product must and should do and it sets the criteria by which the product is either allowed to continue on its development or be killed off during quality control reviews. It should also be a useful and helpful guide to the design team to ensure that nothing is “left out” of the product' s design during its development.

一个有效的机会特性包含该拟议新产品的两个方面：

An effective opportunity specification covers two aspects of the proposed new product：

(1) 它描述了由拟议的新产品所呈现的商业机会，主要关注的是该产品的核心利益主张。

(1) It describes the business opportunity presented by the proposed new product，focusing on the product' s core benefit proposition.

(2) 它通过计划销售、超过制造成本之上的利润、开发成本和计划盈亏平衡点(在该点开发成本将得到弥补)，提供该商业机会的合理性。

(2) It provides a justification of that business opportunity in terms of projected sales，margin over manufacturing costs，development costs and projected break-even point(at which development costs will be recovered).

设计规范草案格式的实例

An example of format for a draft design specification

提议的产品：土豆去皮刀 Proposed product：Potato peeler	设计者：约翰 Designer：John	
日期：1999．01．15 Date：Jan．15，1999		
性能要求 Performance requirement	设计要求 Design requirement	设计规范 Design Specification
必须要能把土豆上的凹眼挖掉 Must remove the eyes from potatoes	必须要有一个圆凿子 Must have a gouge 必须设计成不需紧握手把就可以用此圆凿子 Must be designed so that the gouge can be used without having to grip the handle	圆凿子必须靠近手把 Gouge must be positioned close to the handle 圆凿子应可以在不改变去皮刀上的手把时使用 Gouge should be used without having to change the grip on the peeler

对设计团队来讲设计规范也应该是一份实用而有益的指南，以保证在产品开发过程中产品设计不会有任何“遗漏”

It should also be a useful and helpful guide to the design team to ensure that nothing is “left out” of the product’s design during development

拓展阅读

Extend and reading

Plastech Ltd.：potato peeler

Plastech’s development and innovation of the potato peeler is worth noticed example in product planning. Plastech Ltd. is a small company with 170 employees manufacturing a range of small plastic domestic goods, and most are small kitchen products. Before development of the potato peeler, they generated a profit of 900000 pound sterling on a turnover of 8 million. The management of the company decided to move to produce higher value products to add value by design.

Their product planning process is: decide the development strategy (Remain in the value-for-money side of their business and not to produce luxury end of the market)→analyze 450 different competing products→carry out qualitative and quantitative market research, to find the customer desires and needs→find the new products variety worth to be developed→analyze design specification of existing 43 potato peelers→modify the potato peeler→financial feasibility and cost-investment return period→justification of the opportunity→develop the new potato peeler→the Plastech's new potato peeler.

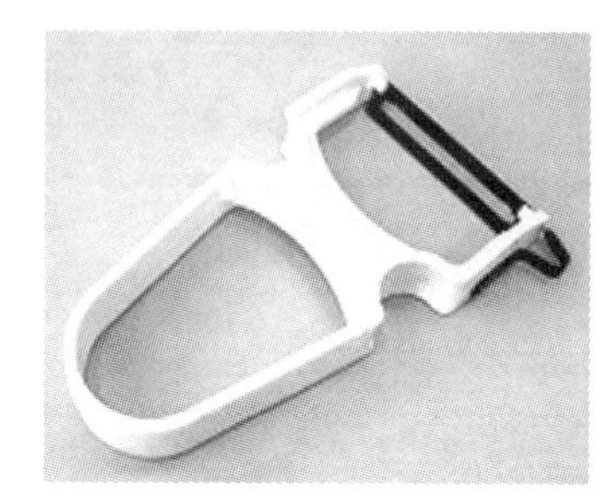

	R&D	Innovation Design	Time to market	Production engineering	Technical marketing	Patents
Pioneering	×××	×××	××	××	×××	×××
Responsive	×	×××	×××	××		×
Traditional					×××	
Dependent					×××	

Proposed product: Potato peeler		Designer: John
Information from: Lan (marketing)		Date: Jan. 151999
Product requirement	Demand or wish	Type of requirement	Basic/Performance/Excitement
Must look hygienic	Demand	Market	Basic
Must feel comfortable to hold	Demand	Market	Performance
Should be super-sharp	Wish	Market	Excitement

Proposed product: Potato peeler		Design: John
Date: Jan. 15, 1999		
Performance requirement	Design requirement	Design specification
Must remove the eyes from potatos	Must have a gouge. Must be designed so that the gouge can be used without having to grip the handle	Gouge must be positioned close to the handle. Gouge should be used without having to change the grip on the peeler

第14课 | 市场需求调查
Lesson 14 | Market needs research

词汇和短语
Words and phrases

market needs research　市场需求调查

product competing strength　产品竞争力

functional speciality　功能特点

product survey　产品调研

sales survey　销售调查

competing product survey　竞争产品调研

product opportunity　产品机会

library research　图书馆(资料)研究

qualitative survey　定性调查

quantitative survey　定量调查

target market　目标市场

questionnaire　问卷，调查表

词汇联想与记忆
Association and memory of words

compete	*vi.*	比赛，竞争
survey	*n.*	测量，调查
	vt.	调查(收入，民意等)，测量
speciality	*n.*	特性，特质，专业，特殊性

qualitative *adj.* 性质上的，定性的

quantitative *adj.* 数量的，定量的

target *n.* 目标，对象，靶子

关键词汇和概念

Key words and concepts

市场需求调查是识别、确定和判断产品机会的基础，是了解需求并使产品开发成功的基点。

Market needs research is fundamental to identifying, specifying and justifying a product opportunity, and a basic point to understand the needs and to make the product development success.

课文

Text

市场需求调查

Market needs research

产品竞争力的关键是产品能否给消费者带来使用上的最大便利和精神上的满足。为使所设计的产品能艺术地反映出功能特点的形象从而使产品开发成功，必须要站在为使用者服务的基点上，从市场调研开始。这类调研可以分为产品调研、销售调研和竞争产品调研。通过市场调研了解消费者的需求绝对是识别、确定和判断产品机会的基础。

The key of the product competing strength is if the product can offer the greatest facilitation and the spirit satisfaction to the customer. To design a product with an image reflecting the functional speciality artistically and to make the product development success, a market needs research is needed based on the point of serving the users. These market needs research includes product survey, sales survey and competing product survey. Understanding the needs of customers through market research is absolutely fundamental to identifying, specifying and justifying a product opportunity.

市场需求调研可以以四个信息来源为基础进行：

Market needs research can be based on four main sources of information:

- 内部的市场信息
- In-house market intelligence
- 图书馆资料研究
- Library research

- 定性市场调查
- Qualitative market surveys
- 定量市场调查
- Quantitative market surveys

在熟悉市场中运行的公司的主要资产之一是其对市场的了解，这一点要在开发机会特性时发挥到极致。公司的销售团队，或者对产品进行维护和修理的服务人员，他们可能对此更为了解。

One of the main assets of a company operating in a familiar market is its knowledge of that market，this must be used full in developing an opportunity specification. The company sales force or the service personnel who maintain and repair products may have a much better understanding.

可以用各种不同的方法将这些信息的精华提取出来：正规的访谈或会议，通过电话、邮件及非正规讨论进行的调研，仔细准备调查问卷，或者请消费者填写“愿望清单”等。工厂的销售记录也能对消费者需求提供有用的提示。但是不能期望公司的信息来源能给出消费者需求的全面描述。当你研究公司的信息来源时，要注意信息的时效性等限制，从而对信息进行加权处理。

This information can be distilled in a variety of ways：formal interview or meetings，survey through telephones，mails，informal discussions，carefully prepared questionnaires，or asking for filling a “wish list” etc.. Company sales records can also provide useful pointers to customer needs. However，company sources of information cannot be expected to provide a complete description of customer needs. When you research company sources you must be aware of limitations such as time limits and weigh up the information accordingly.

图书馆资料的研究并不是指真正在图书馆里进行的研究，而是泛指对所有机构出版的资料的研究。研究机构公开发布的报告也能提供很有价值的信息。

Library research does not necessary mean “research done in a library”. Rather，it should be interpreted more liberally to mean research using organizations published information. Published reports by research organizations can provide a gild mine of information.

对于大的产品开发项目，市场的定性研究和定量研究两者都要进行。定性研究仅是一种探索性的、粗略的判断，目的是探寻消费者的感知和需求，也就是搜寻消费者对需求的判断和看法，以及了解目前的产品是否满足他们这种需求。定性研究可以对范围更宽广的问题给出结果，并就消费者对市场上产品的感觉进行深度的研究。定量研究的调查对象要选择目标市场群体中统计学上有代表性的消费者作样本。这种研究可以面对面或通过电话访问进行，严格坚持按精心设计的问卷进行调查。

AMADOR WATER AGENCY

AMADOR CANAL AND IONE PIPELINE

CUSTOMER QUESTIONNAIRE

CUSTOMER INFORMATION

Customer Name(s): **Account #:**

Service Address:

Legal Owner(s): **APN:** **Phone #:**

QUESTIONNAIRE

Please complete and return this questionnaire form using the envelope provided.

1. Who is completing this questionnaire? ☐ Legal Owner ☐ Renter ☐ Other (Print name and relationship)

2. Is there a residence or business on the property? ☐ Yes ☐ No

If more than one building (residence/business) is on the property, please contact Tammy Hebrard at (209)257-5205 for additional questionnaire forms.

3. The above-referenced property is served water from the following source(s): (Please mark all that apply)

☐ Treated water from the Amador Water Agency.

☐ Store-bought/delivered, bottled water. Store/Company: ______

☐ Private well.

4. Is the raw water used for agricultural purposes only? ☐ Yes (skip to question #8) ☐ No

5. Is the raw water used in your home or business? ☐ Yes ☐ No (skip to question #8)

6. Is the raw water treated in some way? ☐ Yes ☐ No (skip to question #8)

If "yes", please explain what type of treatment or filtration system is used. (Please include Manufacturer and Model#)

7. What is the raw water used for: (check all that apply)

☐ drinking ☐ cooking ☐ personal hygiene (showering, bathing, or brushing teeth)

☐ dishwashing/dishwasher ☐ laundry

8. What source(s) of water do you use for drinking water? ______

CERTIFICATION

I/we hereby certify that I/we are the ☐ Owner(s) ☐ Renter(s) ☐ Other of the above-identified parcel, and the forgoing is true and correct to the best of my/our knowledge.

Signature ______ Signature ______

Print Name ______ Print Name ______

Phone Number Date

调查问卷示例一

Example one of questionnaices

University of Fort Hare
Together in Excellence

Trainee / Customer Questionnaire

Please take 2 minutes of you time to fill in the following questionnaire in order to provide us with feedback to ensure our courses and trainers are of the highest caliber! We greatly appreciate your input.

Course Attended: Trainer:

1. How would you rate the impact your training has had on your performance in the workplace?

   Excellent ○ **Very Good** ○ **Average** ○ **Poor** ○

2. If you have made telephone enquiries to PFSA, how have you found the service?

   Excellent ○ **Very Good** ○ **Average** ○ **Poor** ○

3. How would you rate overall service at the PFSA?

   Excellent ○ **Very Good** ○ **Average** ○ **Poor** ○

4. How would you rate the Facilitators/Trainers at the PFSA?

   Excellent ○ **Very Good** ○ **Average** ○ **Poor** ○

5. If you have any training queries, how were they addressed?

   Excellent ○ **Very Good** ○ **Average** ○ **Poor** ○

6. How would you rate the effectivesness of the PFSA communication? Newletter, heard us on radio, etc.

   Excellent ○ **Very Good** ○ **Average** ○ **Poor** ○

7. General Comments:

PFSA
excellence in public financial management

调查问卷示例二
Example two of questionnaices

For large product development projects, both qualitative and quantitative researches will be used. Qualitative research is exploratory and largely judgmental, aimed to explore the perceptions and needs of customers, that is you are seeking customer's judgments and opinions on their needs and how these needs are satisfied by current products. Qualitative research can give coverage of a wide breadth of issues and can study in-depth customer's perceptions of products in a market. Quantitative research is to survey a statistically representative sample of customers from a target market group. The research is conducted either face-to-face or by telephone and sticks rigidly to a careful devised questionnaire.

拓展阅读

Extend and reading

Technology eats itself

Designers put a whole lot of work into the technical details of the functioning of their products. Yet the world as we know it is so darn complex, problems can arise easily, especially when your product has to interface with other parts of the built environment.

One of the most frightening examples(from a corporate standpoint)was the recent discovery that the most common bike lock, made by Kryptonite, could be picked in about ten seconds with a Bic ballpoint pen. The idea is that the circular opening of the barrel lock was identical in size to the barrel of the most common Bic pen, and that the plastic would moosh enough to push the pins down and sideways in such a way that the pins would find their seats automatically. A video of this feat was released, and within a few days it was common knowledge in bike circles that your lock was useless. During the course of a few days, a technology that had been robust enough to thwart the most experienced bike thieves on the planet could be bypassed by a kid with a pen.

What's interesting is that Kryptonite had been working for years to build a lock that could stand up to be "incompatible" with hacksaws, prybars, car jacks, and anything else short of liquid nitrogen or an acetylene torch. Yet a product was discovered in an entirely unrelated field that had a disastrous compatibility with the locks.

The more products that are released into this world, the more chance that product Q will unexpectedly have a problem with product X. There are some parallels in the natural world, such as the huge numbers of chemicals that plants create to defend themselves from the huge number of chemicals that other plants create, etc.. Speaking of chemicals, don't ever mix ammonia products with bleach—it'll create a toxic gas that's way worse than either of the two original products. Yet another unexpected breakdown in our technology.

Back in the mechanical world, another disastrous interface has yet to be overcome is the height of SUV bumpers. Most bumpers(this is changing slowly)on SUVs are higher than the side rail of a passenger car or minivan. When the smaller car is hit in the side by the SUV, the SUV rides over the internal structure of the car and into the passenger area, increasing the chance that the passengers will be killed. I got a graphic demonstration if this when a car I was riding in was hit on the driver's side by a Ford Explorer(circa 1995). The driver was injured much worse than if the two vehicles were of equal height. Looking at the wreck later it was apparent that the SUV had avoided the bottom rail of the car altogether, and crushed the left-side occupants over a foot and a half. A lower bumper would probably have lessened the injuries. Maybe.

We're not going to get into issues of negligence or fault here, but some of these examples illustrate that even when a product is perfect within the scope of its application, there's a whole world of objects out there that can make things go wrong.

第15课 | 产品企划
Lesson 15 | Product planning

词汇和短语
Words and phrases

product development strategy　产品开发战略

product innovation　产品创新

business success　商业成功

free fall　自由落体

tangible result　有形结果

(be)conducted to　旨在

viability　生存能力，生存性

market pull　市场吸引

technology push　技术推动

词汇联想与记忆
Association and memory of words

strategy　*n.*　策略，战略

innovation　*n.*　改革，创新

tangible　*adj.*　切实的

conduction　*n.*　传导性

viability　*n.*　生存能力，发育能力

push　*n.*　推，推动；奋发，进取心

vt. 推，推动，推行

关键词汇和概念

Key words and concepts

产品企划包括确定机会，进行市场调研，分析竞争产品，提出新产品方案，拟定机会特性和设计规范。产品企划是开发新产品中最难的活动之一。

Product planning includes spotting an opportunity, conducting market research, analyzing competing products, proposing a new product and drawing up both an opportunity specification and a design specification. Product planning is one of the most difficult activities in the development of new products.

课文

Text

产品企划

Product planning

产品企划包括确定机会，进行市场调研，分析竞争产品，提出新产品方案，拟定机会特性和设计规范。产品企划是开发新产品中最难的活动之一。

Product planning includes spotting an opportunity, conducting market research, analyzing competing products, proposing a new product and drawing up both an opportunity specification and a design specification. Product planning is one of the most difficult activities in the development of new products.

产品企划始于公司的产品开发战略，终于作出新产品的设计规范。一般来说，产品的开发战略是公司进行创新的方法，它是公司如何将产品的创新转变为商业上的成功的计划。它叙述了公司产品在市场的定位，从而决定了公司寻求开发的新产品种类。

Product planning begins with the company's product development strategy and ends with a design specification for a new product. A product development strategy, in general term, is the company's approach to innovation. It proposes how that company plans to turn product innovation into business success. It describes the positioning of the company's products in the market and thereby determines the sort of new products the company seeks to develop.

设计师把产品开发中的企划阶段称为"自由落体"。你带着无数稍纵即逝的想法闪电般地从空中穿过并设法在这些想法一闪而过以前抓住它。在此自由下落过程中，如果你能系统地和充分地推定出设计规范，你的降落伞就打开了——充满信心地成功软着落，并可直接着手进入产品设计。如果你不能捕捉完全决定了的设计规范，那么，着落将会很艰难。

Designers often describe this part of product development as the period of “free fall”. You are hurtling through space with a myriad of ideas flashing past and you reach out to grasp for them, before they disappear beyond your reach. If during free fall, you are able to derive a design specification systematically and thoroughly, your parachute opens—you land softly, with confidence and ready to move straight into product design. If you fail to get that design specification properly resolved, the landing will be harder.

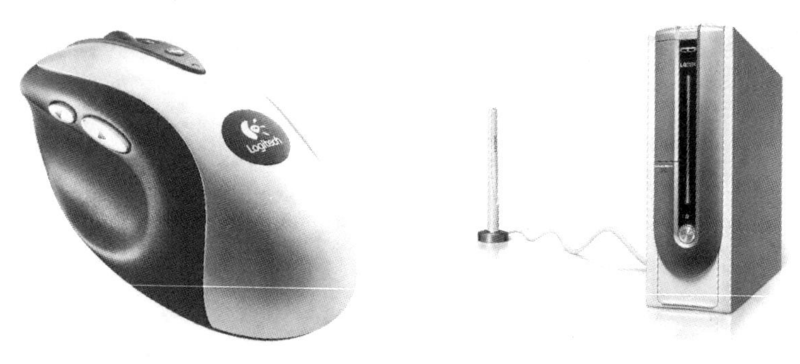

产品企划可分四个主要阶段：①产品开发策略：给出产品计划的大方向并设定目标；②“产品触发器”，启动产品开发的促进因素；③研究和分析机会和限制条件；④提议的新产品最终确认与判定。

Product planning can be seen as falling into four main stages. Firstly, there is the product development strategy: giving a general orientation to product planning and establishing its objectives. Secondly, the “product trigger” that is the stimulus to start the development of a product. Thirdly, the opportunities and constraints are researched and analyzed and finally the proposed new product is specified and justified.

产品企划的有形结果是管理者同意设计该新产品。在这点上，管理者最关心的是新产品和现有产品的差异以及这种差异会给企业带来什么样的机会。而设计师需要的是，说明产品必须具有的特性以及如何将它做出来的详细规范。另一个差别是，管理者往往要设定并坚持一个经营目标，而设计师则希望该规范有一定程度的灵活性。为此，我们需要承担两个层面的义务：对项目商业目标的承诺，以及使新产品的技术特性有更灵活规范的承诺。

The tangible outcome from product planning is getting commitment from management to begin designing the new product. Of most interest to management is how this new product is going to differ from existing products and what business opportunities arise from this differentiation. For the designer, a more detailed specification is needed describing what features the product must have and how can they be made. Another difference is that management will want to set the business objectives and stick to them. The designer, on the other hand, will want a degree of flexibility in the specification. What we need, therefore, is two levels of commitment. Firstly, commitment to the business objectives for the project, and secondly, commitment must be made to a more flexible specification of the specific technical features of the new product.

一般来说，研究和分析是产品企划中很耗时的部分。对所有产品企划的指导原则是，若一个产品机会确认了产品在商业上的生存能力，并证明与公司的产品开发战略一致时，该产品企划就是满意的了。

Research and analysis is generally the time-consuming part of product planning. The guiding principle in all product planning is that a product opportunity is satisfactory when it confirms the commercially viability of the product and demonstrates consistency with the company's product development strategy.

有两大类产品触发器：市场牵引和技术推动。市场牵引是指市场对你公司目前尚未提供的某一产品或者产品特性的需求。技术推动是指，可得到的一种新技术能对产品的创新创造机会。这种新技术可以是一种新材料、新的制造工艺或新的设计概念。

Product triggers fall into two main categories: market pull and technology push. Market pull refers to the demand by the market for a product or product features not currently offered by your company. Technology push refers to the available of a new technology creating an opportunity for product innovation. This new technology could be a new material, a new manufacturing process or a new design concept.

要牢记两件事：①产品企划做得越好，产品商业成功的机会越多。②产品企划上花的时间越多，后来产品开发时可节省的时间也越多。

There are two important things to remember: ①The better the product planning, the more chance the product has of commercial success. ②The more time spent in product planning the more time will be saved later in product development.

拓展阅读

Extend and reading

Revenge of the right brain

Logical and precise, left-brain thinking gave us the Information Age. Now comes the

Conceptual Age—ruled by artistry, empathy, and emotion.

When I was a kid—growing up in a middle-class family, in the middle of America, in the middle of the 1970s—parents dished out a familiar plate of advice to their children: get good grades, go to college, and pursue a profession that offers a decent standard of living and perhaps a dollop of prestige. If you were good at math and science, become a doctor. If you were better at English and history, become a lawyer. If blood grossed you out and your verbal skills needed work, become an accountant. Later, as computers appeared on desktops and CEOs on magazine covers, the youngsters who were really good at math and science chose high tech, while others flocked to business school, thinking that success was spelled MBA.

Tax attorneys, Radiologists, Financial analysts, Software engineers, management guru Peter Drucker gave this cadre of professionals an enduring, if somewhat wonky, name: knowledge workers. These are, he wrote, "people who get paid for putting to work what one learns in school rather than for their physical strength or manual skill." What distinguished members of this group and enabled them to reap society's greatest rewards, was their "ability to acquire and to apply theoretical and analytic knowledge." And any of us could join their ranks. All we had to do was study hard and play by the rules of the meritocratic regime. That was the path to professional success and personal fulfillment.

But a funny thing happened while we were pressing our noses to the grindstone: the world changed. The future no longer belongs to people who can reason with computer-like logic, speed, and precision. It belongs to a different kind of person with a different kind of mind. Today—amid the uncertainties of an economy that has gone from boom to bust to blah—there's a metaphor that explains what's going on. And it's right inside our heads.

Scientists have long known that a neurological Mason-Dixon line cleaves our brains into two regions—the left and right hemispheres. But in the last 10 years, thanks in part to advances in functional magnetic resonance imaging, researchers have begun to identify more precisely how the two sides divide responsibilities. The left hemisphere handles sequence, literalness, and analysis. The right hemisphere, meanwhile, takes care of context, emotional expression, and synthesis. Of course, the human brain, with its 100 billion cells forging 1 quadrillion connections, is breathtakingly complex. The two hemispheres work in concert, and we enlist both sides for nearly everything we do. But the structure of our brains can help explain the contours of our times.

Until recently, the abilities that led to success in school, work, and business were characteristic of the left hemisphere. They were the sorts of linear, logical, analytical talents measured by SATs and deployed by CPAs. Today, those capabilities are still necessary. But they're no longer sufficient. In a world upended by outsourcing, deluged with data, and choked with choices, the abilities that matter most are now closer in spirit to the specialties of the right hemisphere—artistry, empathy, seeing the big picture, and pursuing the transcendent.

Beneath the nervous clatter of our half-completed decade stirs a slow but seismic shift. The Information Age we all prepared for is ending. Rising in its place is what We call the Conceptual Age, an era in which mastery of abilitics that we've often overlooked and undervalued marks the fault line between who gets ahead and who falls behind.

To some of you, this shift—from an economy built on the logical, sequential abilities of the Information Age to an economy built on the inventive, empathic abilities of the Conceptual Age—sounds delightful. "You had me at hello!" I can hear the painters and nurses exulting. But to others, this sounds like a crock. "Prove it!" I hear the programmers and lawyers demanding.

OK! To convince you, I'll explain the reasons for this shift, using the mechanistic language of cause and effect.

第16课 | 设计怪才路依吉·柯拉尼
Lesson 16 | Monster-designer Luigi Colani

词汇和短语
Words and phrases

monster 巨人，怪物

controversial 有争议的

critic 批评家，评论家

entertainer 演艺家，艺员

genius 天才

philosopher 哲学家，哲人

aerodynamics 空气动力学

catamaran 双体船

optical frames 眼镜架

ergonomic 人机工程学，人体工程学

dominant 占优势，支配地位，有统治权

provoker 煽风点火者

guru 宗教老师，领袖，头头

undulating 波浪形的

contour 轮廓，等高线

trustee 托管人，保管人

straying 迷路的，离群的

functionality 功能性

renown 名声，声望

translator 翻译者，解释者

interpreter 翻译者，口译人员

modesty 谦逊，虚心

aeon 万古，永世，十亿年，漫长岁月

distinct 独特的，截然不同的

bio-dynamics 生物动力学

imitate 模仿，仿造

词汇联想与记忆

Association and memory of words

critical *adj.* 评论的，鉴定的，批评的，危急的，临界的

entertainment *n.* 款待，娱乐，娱乐表演

philosophy *n.* 哲学，哲学体系，达观，冷静

dynamics *n.* 动力学

关键词汇和概念

Key words and concepts

路依吉·柯拉尼确实是我们时代最著名的设计师之一，也是最有争议的设计师之一。"我只是个对自然的诠译者"，柯拉尼自己说。

Luigi Colani is certainly one of the most famous designers of our time, yet one of the most controversial as well. "I'm only a translator of nature", Colani says himself.

课文

Text

设计怪才路依吉·柯拉尼

Monster-designer Luigi Colani

路依吉·柯拉尼确实是我们时代最著名的设计师之一，也是最有争议的设计师之一。有人把他看作一个职业批评家或者一个设计演艺家，但是也有人几乎把他当作天才和哲人来崇拜。

Luigi Colani is certainly one of the most famous designers of our time, yet one of the most controversial as well. There are the ones who see him as a professional critic or a design entertainer, but there are also those who almost worship him as a genius and a philosopher.

路依吉·柯拉尼无疑是我们时代最著名的设计师之一
Luigi Colani is certainly one of the famous designers of our time

路依吉·柯拉尼1928年生于柏林。他先后学习过雕塑和空气动力学。然后在1953年，他成为加利福尼亚州一家公司的新材料项目组的头头。1958年他设计的竞速双体船获得了成功。自此，他设计过了许多东西。

Luigi Colani was born in Berlin in 1928. He studied of sculpture and then aerodynamics. After study, he became the head of new materials project group at a company in California in 1953. In 1958, racing catamaran designed by him gets success. From this, he has been designed many objects.

柯拉尼设计的佳能T90
Canon T90 was designed by Colani

他不仅设计了诸如轿车、摩托车和飞机这样的产品，还设计了许多其他影响我们日常生活的普通物品。这些物品包括体育器械、家庭用品、家具、盥洗室用具、照相机和其他电子器材、计算机、宝石、眼睛架等。只举一个例子，路依吉·柯拉尼教授设计了佳能T-90照相机。

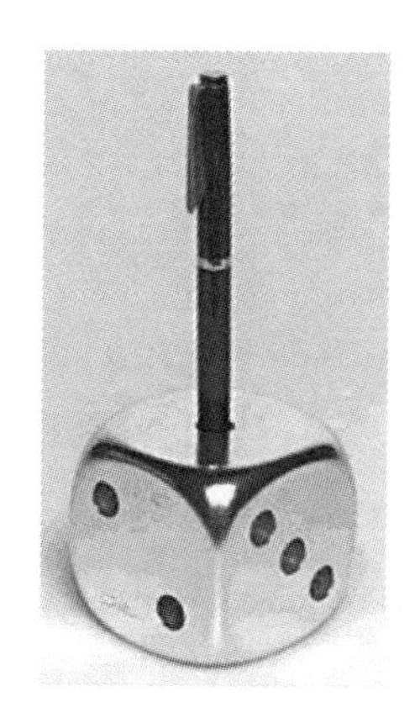

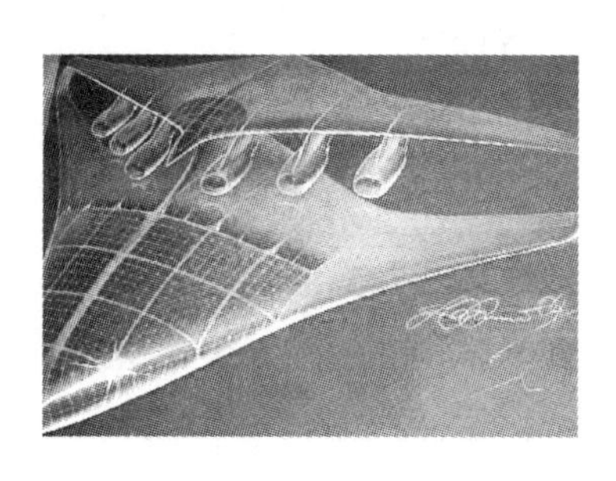

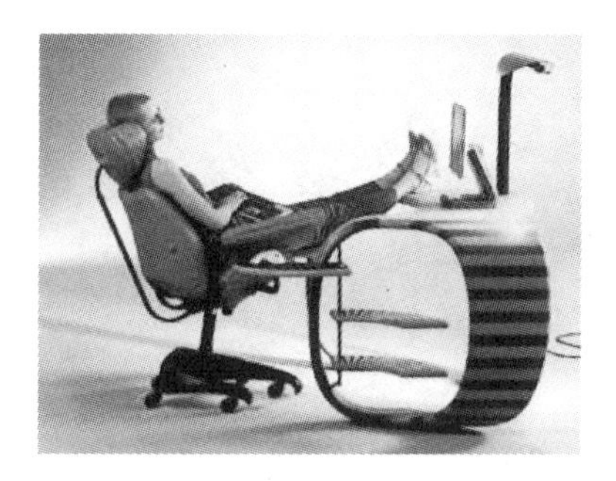

He has not only designed products such as cars, motorcycles and aircraft but also many other common items that affect us in our daily lives. These products include sport equipment, household goods, furniture, bathroom appliances, cameras and other electronic equipments, computers, jewellery, optical frames and so on. Just to mention one example, Prof. Luigi Colani has designed the Canon T-90 camera.

这款照相机不仅赢得了1987年度的照相机奖，而且确定了照相机人机工程学设计的趋势，这个趋势一直风行至今。他为马自达公司设计了马自达MX-5型汽车，一款圆球形的汽车，这也设定了未来汽车设计的趋势。

This camera not only won the camera of the year 1987 award, but also set the trend for ergonomic camera design which is dominant today. And for Mazda, he designed the Mazda MX-5, a round shaped car design which has set the trend for future car designs.

在40多年的创造活动中，路依吉·柯拉尼赋予范围广泛的产品以时尚和造型从而带给整个工业许多推进，产生了国际性的影响，他永远是一个驱动力。

In more than 40 years of creative activity, Luigi Colani has always been an driving force, given many impulses and influenced internationally entire industries by fashioning and shaping a wide range of products.

“地球是圆的，天空所有的物体也是圆的，它们都在圆形或者椭圆形的轨道上运行。”这种彼此环绕旋转的球形世界的相同图形一直追随我们直至微观世界。“为什么我还要加入到想让每个东西都有棱角的迷路人群中去呢？我正在追赶伽利略的哲学：我的世界也是圆的。”

“The earth is round, all the heavenly bodies are round; they all move on round or elliptical orbits.” This same image of globe-shaped worlds orbiting around each other follows us right down to the microcosmos. “Why should I join the straying mass who want to make everything angular? I am going to pursue Galilei's philosophy: my world is also round.”

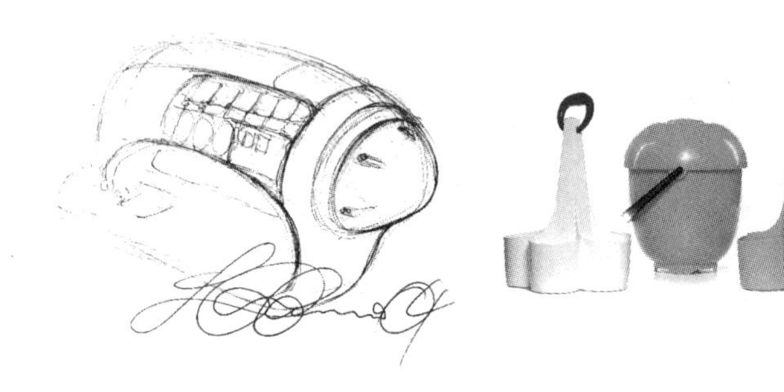

柯拉尼完美和实用的设计，显现出了他在人机工程学、空气动力学和功能性的手法方面的创造性及全新的方法。柯拉尼是个超前于他的时代的、有国际声望的伟大设计师。

Colani's perfect and practical designs appear his creativity and entirely novel approach in terms of ergonomics, aerodynamics and functionalities. Colani is a great designer of international renown well ahead of his time.

但他至今仍然谦虚地把自己看做是个自然的诠译者。柯拉尼近距离地观察自然，自然本身在漫长的岁月里的发展已经产生了无数尽善尽美的奇迹，而他则对此进行了分析、适应、调整和充实。通过显示和传达他的产品不同于其竞争者的独特之处，他面对挑战！

Yet, in his modesty, he sees himself merely as a interpreter of nature. Colani is closely observing nature where development over aeons has produced uncountable wonders of perfection and beauty, and he further analyzes, adapts, adjusts and implements. He is facing the challenge by showing and communicating what distincts his products from those of the competitors!

全新的解决方案和/或创新，常常是通过自定的准则而发展起来的，有时会被看做是投机；但是随后就会发现，在实际应用中非常优异，在实际体验中也是令人心悦诚服的。

Entirely novel solutions and/or creations are often developed through self-imposed criteria and may at times be seen as speculative, but are subsequently found to be excellent in practical application and convincing in positive experience.

自然是出发点。这是路依吉·柯拉尼教授的生物动力学哲学里的中心概念。柯拉尼，从自然得到许多灵感，强调他的设计是基于自然界生物的形态组成的。“我所做的不比模仿自然显露给我的真相更多。”

Nature is the starting point. This is the central concept in Prof. Luigi Colani's philosophy of bio-dynamics. Colani, who gains much of his inspiration from nature, stresses that his designs consist of shapes based on the creations of nature: “I do not more than imitate the truths revealed to me by nature!”

拓展阅读

Extend and reading

Famous Designers: Luigi Colani, Michael Graves, Philippe Starck

Luigi Colani

Provoker, challenger, aesthete, design-guru, creator of practical forms and perfect shapes, promotor of round and undulating patterns, professor of ergonomics, style and contours, trustee of the philosopher's stone.

The earth is round, all the heavenly bodies are round; they all move on round or elliptical orbits.

“I'm only a translator of nature” Colani says.

“Even objects existing in the natural world are made up of curves.”

This same image of circular globe-shaped mini worlds orbiting around each other follows us right down to the microcosmos. We are even aroused by round forms in species of propagation related eroticism. “Why should I join the straying mass who want to make everything angular? I am going to pursue Galilei's philosophy: my world is also round.”

Michael Graves

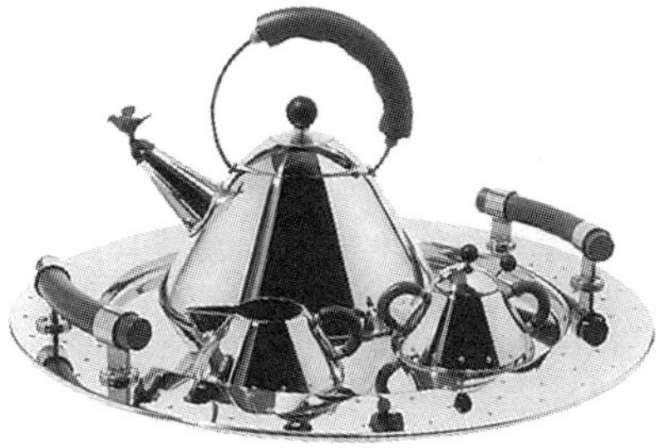

He teaches architecture in Princeton since 1962. He created the best-selling 9093 kettle for Alessi.

"As early as January 1980, during our first visit, Michael Graves told us that from then on he would be spending at least half of his time on design.

It was a definitive statement which, moreover, corresponded to his great potential in this field. Michael does not like theory even though he once confessed his desire to develop an 'American' design.

In any case, between the 1980s and 1990s he showed an incredible ability to tune in to the tastes of even the average public. He has shown he can bewitch the public like only very few of the designers with whom I've worked."

Philippe Starck

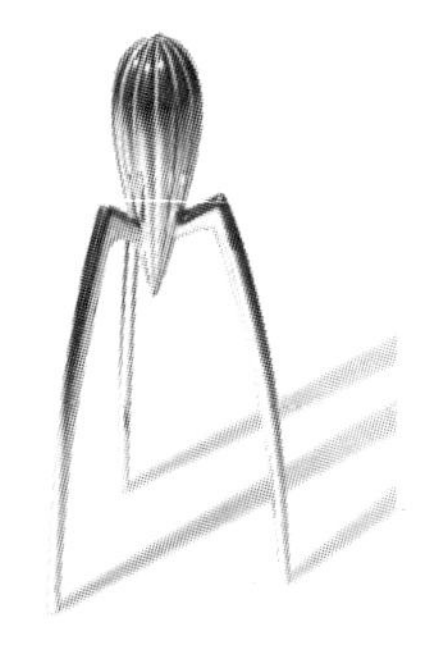

Born in Paris in 1949, he is one of the most original and creative designers of our time. He has obtained many important acknowledgements, he considers himself as "a Japanese architect, a American art director, a German industrial designer, a French artistic director, an Italian furniture designer".

"Starck began working with me in 1986, on a French design project.

He is a living example of my dream: design, real design, is always highly charged with innovation towards the world of manufacturing and trade, bringing results that need no longer be justified solely on a technological or balance sheet level.

A true work of design must move people, convey feelings, bring back memories, surprise, transgress… in sum, it has to be poetic. Design is one of the most apt poetic form of expression of our day.

And I know that this great visionary still has plenty of surprises up his sleeve, despite his threats to retire!"

第5单元
Unit 5

第17课 | 产品开发策略
Lesson 17 | Strategy for product development

词汇和短语
Words and phrases

strategy for product development　产品开发策略

opportunism　机会主义者

with regard to　关于

to some extent　在某种程度上

strategic planning　战略性规划

a succession of　一连串

pioneering strategy　先锋型策略

pro-active strategy　前摄型策略

innovation culture　创新文化

patent protection　专利保护

monopoly profit　垄断利润

responsive strategy　响应型策略

traditional strategy　传统型策略

static market　静态市场

dependent strategy　依赖型策略

priority　优先次序

time to market　上市时间

business strategy　经营策略

risk management　风险管理

词汇联想与记忆

Association and memory of words

extent *n.* 广度，宽度，长度；范围，程度

strategics *n.* 兵法，战略学

patent *n.* 专利权，执照，专利品

adj. 特许的，专利的

vt. 取得……的专利权

profit *n.* 利润，益处，得益

vi. 得益，利用

关键词汇和概念

Key words and concepts

产品开发策略由企业的经营策略决定。它提供指导方针来决定公司寻求发展的新产品类型。

Strategy for product development is determined by enterprise's business strategy. The strategy provides the guiding light to determine the type of new products the company seeks to develop.

课文

Text

产品开发策略

Strategy for product development

一些公司经常说，他们对待创新是持机会主义态度的。这在一定程度上说明，他们没有抓住战略性规划的要领。作为机会主义，就是说不能控制局面，或者至少不能预知将要作用于你的压力及其影响的情况下所做的事情。机会主义是“救火队”式的管理形式，这时经理的主要任务就是在危机四伏中驾驭公司。但是公司是不应该依靠机会主义者来获得商业成功的。不同的策略要求公司有不同功能的人员和资源的投入。与创新有关的策略可以有以下四种基本类型：

Very often, companies say they are opportunistic with regard to innovation. To some extent, this means they are missing the point of strategic planning. Being opportunistic is what you do when you are unable to control or at least predict the pressures and influences which affect you. Opportunistic is a form of "fire brigade" management in which the manager's main task is to steer the company through a succession of crises. But companies should not depend upon opportunism for commercial success. Different strategies require different investments of people and resources in different functions with a company. Strategies with regard to innovation can be categorized as four basic types as listed below:

(1) 先锋型策略。其目标是争当技术和市场上的领导者。这些公司非常依赖于研究和开发，领先于竞争者给市场引入激进与渐进两方面的创新。

(1) Pioneering strategies. They aim to give technical and market leadership. They depend heavily upon research and development to introduce both radical and incremental innovation to the market ahead of their competitors.

先锋型策略是一种前摄型策略，对投资的回报持长期的观点。它在很大程度上依赖于公司内部有效的创新文化。十分重视专利保护，因为公司通常需要用实实在在的垄断利润去支付用于产品开发及不可避免的产品失败方面的投资。

A pioneering strategy is pro-active and has a long-term perspective on return on investment. It is highly dependent upon an effective innovation culture within the company. Considerable importance is attached to patent protection, since substaintial monopoly profits will generally be required to pay for investment in product development and the inevitable product failures.

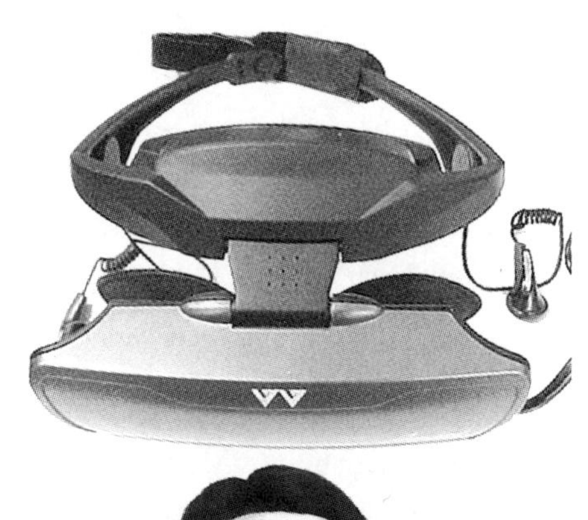

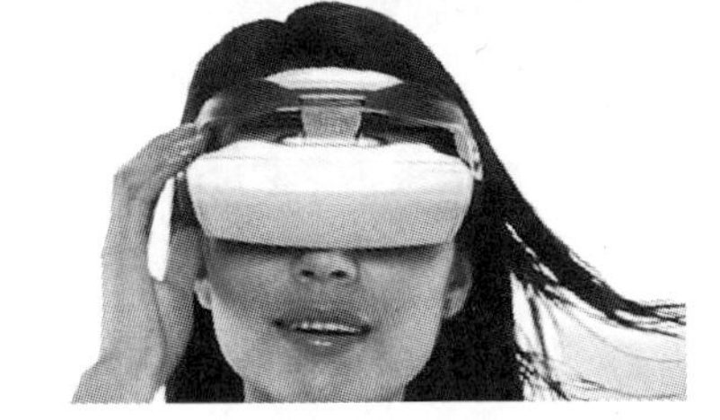

他们以技术和市场领导地位为目标(进行设计)
They aim to give technical and market leadship

(2) 响应型策略。其目标是对先锋型竞争者作出响应，但是有意地让其他公司去承担产品开发的风险并去开拓新的市场。依靠对市场的快速跟进，并常常努力去改进先锋型的产品(第二位，但是更好的策略)。

(2) Responsive strategies. They aim to respond to the pioneering competitors but deliberately let others take the new product development risks and open up new markets. Depend upon fast times to market and often strive to improve upon pioneering products(a 2-nd-but-better strategy).

(3) 传统型策略。采用这种策略的公司大多长期经营在产品范围很稳定的巨大静态市场上，此时市场需求很少或没有变化。即便创新，通常也仅限于产品的微小变化以便降低成本，便于生产或提高产品的可靠性。

他们以向先锋地位的竞争者作出反应为目标(进行设计)
They aim to respond to the pioneering competitors

(3) Traditional strategies. They are adopted by companies operating in a largely static market with a largely static range of products, where there is little or no market demand for change. Innovation normally limited to minor product changes in order to reduce the costs, ease production or increase product reliability.

(4) 依赖型策略。公司将创新依赖于其母公司或其消费者时就采取这种策略。依赖型公司的内部创新一般仅限于工艺过程创新。

(4) Dependent strategies. They are the result of companies depending upon their parent company or their customers for innovation. In-house innovation in dependent companies is usually limited to process innovation.

公司必须要使其组织、管理和资源配置适应其经营策略。策略是公司整个结构和管理的基础，并提供指导方针来决定公司寻求发展的新产品的类型。公司的使命可以是、也常常应该是理想化的。理想可以鼓舞和激励员工，但所追求战略目标的方式必须是现实的。朝向最终战略规划的每一步都必须慎重考虑，并进行商业上的验证。

The company must become organized, managed and resourced to suit its business strategy. Strategy forms the basis for the entire structure and management of the company and provides the guiding light to determine the type of new products the company seeks to develop. A company's mission can, and often should, be idealistic. Ideal can be inspiring and motivating to staff. But the way in which the strategic goals are pursued must be realistic. Every step towards the ultimate strategic plan must be carefully considered and commercially justified.

拓展阅读

Extend and reading

DE-BONO 6 thinking hats for evaluating the idea

White hat：information	What information is needed?
	What information is available?
	What information is missing?
	How do we get the information we need?
Black hat：logical negative	Caution
	Risk assessment
	Criticality
Yellow hat：logical positive：	Benefits
	Values
	Value sensitivity
	How can we make something work?
Green hat：creative	New ideas
	Alternatives
	Possibilities
	Provocation—lateral thinking
Red hat：allows to put forward feeling	Feelings
	Intuition
	Emotions
Blue hat：process of thinking	How do we define the solution?
	Decisions
	What are the aims?

What do De-Bono 6 thinking hats do：

By directing every one's thoughts to combine in parallel to different modes of thinking，convert potential conflict into co-operation and separation of evaluative modes of thinking，allow for constructive decision making and evaluation of ideas.

Design in virtuality

If you'll remember a few weeks ago，we had a bit on the growing opportunities for integrating virtual information into the real world through design. But interestingly，with the recent surge of massively multi-player online games, there has been an explosion of the opposite. Designers are turning to the virtual world as an outlet, and sometimes even an income source for their designs. Will wonders never cease?

Aimee Weber is a fashion designer. Her styles are straight off the streets of New York, like the Madonna outfits of the mid 1980s. Talking in a recent interview about her design process:

"Well, it normally starts with me seeing stuff I like in the streets. I also keep a digital camera in my purse, in case I see something cool, like a building front, or a dumpster, or cloth pattern I like. And I just collect all these textures. I have them all over my computer. Then when I see a fashion I like, maybe a chickie with a cool skirt on the subway, something urban chic, I go home."

But, in order to get your hands on her designs, you've got to get virtual. Her clothing and accessories is only available in-game, through her boutique in Second Life, a massively multiplayer online game. And she's not alone. Second Life has a growing number of users who focus on designing elaborate homes, cars, furniture, and even SciFi-type vehicles.

Some of the projects are so elaborate that they almost become "second jobs". Baccara Rhodes, a real life events coordinator recently planned the online wedding of two in-game friends. The elaborate affair took place on a boat, and needed weeks of preparation and hundreds of dollars of real life money. An even bigger production was staged for a birthday by transforming Cayman Island into L Frank Baum's Wizard of Oz. Even the witch with the skywriting was there!

Of course, it's not all fun and games. The Jessie war was a ridiculously bloody conflict which flared for a while in one of the player-killer sections of the simulator. Complete with player built walls and guns, the conflict has finally burned itself out, but who knows what violence is still simmering? Another interesting, but disturbing use of in-game creation is "The White Room", a photoshoot staged inside the "Max Paine Game Engine". The artist created scenes which suggested previous violence that prompted viewers to fit histories to the scenes.

Currently, all this is triviality—it's fun, but it's not really design, is it? Well, in the coming years, as more of our time is spent online, and presumably, using virtual representations of ourselves and others, won't our physical design skills come into play just the same as they do now? As computing becomes more and more a physical analogue, and less and less a 2-D abstraction, physical design, ergonomics, and user understanding will become just as important as they are now in their real-world counterparts. Until then, if you're interested, Second Life probably needs cool looking product. What are you waiting for!

第18课 | 产品功能分析与功能树
Lesson 18 | Product function analysis and function tree

词汇和短语
Words and phrases

product function analysis　产品功能分析

function tree　功能树

function analysis systematic technique(FAST)　功能分析系统技术

customer-oriented　消费者导向的，以消费者为中心的

prime function　主要功能

basic function　基本功能

subsidiary functions　辅助功能，补充功能

(be)irrelevant(to...)　(与……)不相关

词汇联想与记忆
Association and memory of words

orient *n.* 东方，东方诸国

adj. 东方的，上升的，灿烂的

vt. 使朝东，使适应，确定方向

prime *n.* 最初，青春，精华

adj. 主要的，最初的，有青春活力的，最好的，第一流的

v. 预先准备好，[口]让人吃(喝)足，灌注，填装

关键词汇和概念

Key words and concepts

产品功能分析旨在清晰地了解产品各种复杂功能之间的关系。功能树则是一种用树状结构正确有序地、分级地进行功能分析的好方法。

Production function analysis aims to understand clearly relationship of complex product functions. Function tree is a good approach to carry out the function analysis orderly and hierarchically, using a tree structure.

课文

Text

产品功能分析与功能树

Product function analysis and function tree

产品功能分析是系统地分析产品所执行的(使用者感知的)功能的一种方法，也称为“FAST”(即英文 Function Analysis Systematic Technique 的首字母缩写)。在新产品开发中，这是最基本、也可能是最重要的分析技术。对于产品功能分析，所有你需知道的只是产品在使用中是怎样运作的。你必须知道或者能够预知消费者感知到的产品功能，以及消费者如何评价这些功能的相对重要性。这个方法既可用于现有的产品，也可用于仍在设计中的产品。产品功能分析提供了从功能和以消费者为中心的视点对产品的详细理解，并将这种理解在逻辑的和系统的框架中表达出来。其结果可以用来刺激概念的产生。

Production function analysis is a method of systematically analyzing the functions performed by a product(as perceived by the user). Also known as FAST analysis(Function Analysis Systematic Technique), it is the most basic and probably the most important analytical technique in new product development. All you need for product function analysis is know how the product will operate in use. You must know, or be able to predict, the function of the product as perceived by the customer and how the customer rates the relative importance of these functions. It can be applied both to existing products and to those still being designed. Product function analysis provides a detailed understanding of the product from a functional and customer-oriented point of view and presents this understanding in a logical and systematic framework. Its results can be used to stimulate concept generation.

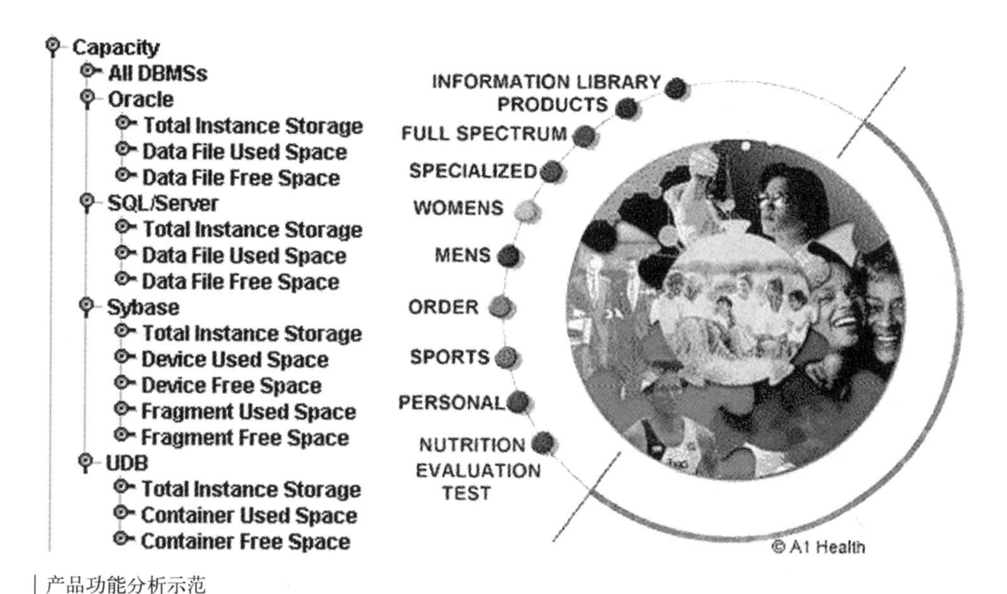

产品功能分析示范
Example of product function analysis

产品功能分析的第一步是用头脑风暴法将产品在消费者眼中会提供的功能全部列出来。最好的方法是将你相信你的产品将执行的功能写在纸条上(贴纸为最理想)。这就是说，要问这个产品是“干”什么功能的，而不是问它“是”什么东西。要从消费者的视点出发做这件事，确认你所写下的功能都是消费者重视的功能。不要认为任何功能是天经地义的。尽量把功能的描述简化到只有2～3个字的“动词—名词”组合。大多数产品至少会有40～60种功能，只有最简单的产品才会有20种以下的功能。

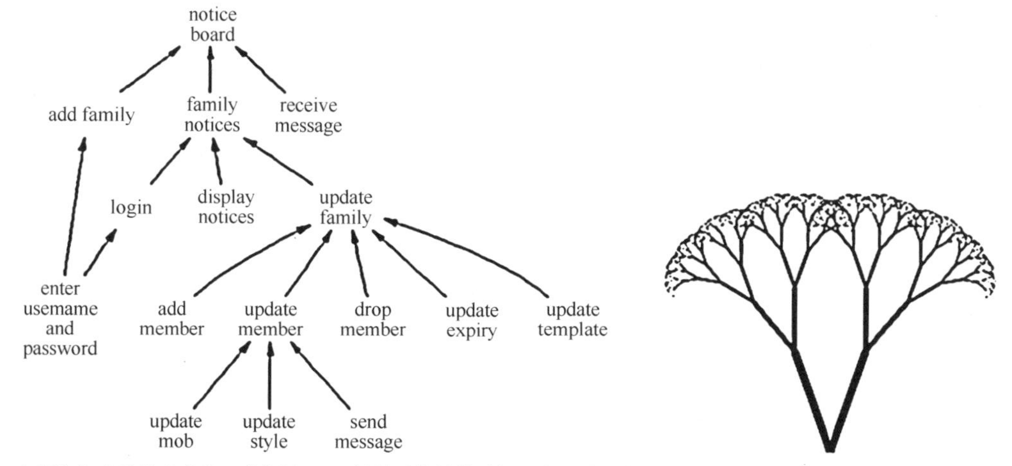

将这些功能整理成为“功能树”。开始做功能树的时候，先要选出产品的主要功能。选定了主要功能之后，其他的功能就按逻辑分组，一层一层地分级排在下面
Arrange these functions into a “function tree”. To start the function tree, select the prime function of the product. Once the prime function has been selected, the other functions are grouped logically and hierarchically under it.

The first step in production function analysis is to brainstorm all the functions the product will serve in the eyes of the customer. The best approach is to write down on individual scraps of paper(post-it notes are ideal)every single function that you believe your product will perform. This means asking what the function that the product "does" rather than what the product "is". Do this from a customer point of view but make sure that you write down all the functions that the customer values. Do not take any function for granted. Try to keep the description of function to two or three words "Verb-noun" combinations. Most products will have at least 40～60 functions and only the simplest products will have fewer than 20.

然后，将这些功能整理成为"功能树"。开始做功能树的时候，先要选出产品的主要功能，即在消费者眼中该产品存在的主要原因。选定了主要功能之后，其他的功能就按逻辑分组，一层一层地分级排在下面。主要功能的下一层应是"基本功能"。基本功能与主要功能有两种方式关联：①它们对主要功能的执行是必要的；②它们是主要功能发生的直接原因。审视整个功能树，问一下"这个功能如何实现？"每一层面上的功能，对上层功能而言应是必要的，并且是上层功能的直接原因。在功能树的底部列出的一串功能，应该是不能再轻易地或逻辑地被进一步细分了。在大多数功能分析中，这些最低层次的功能直接与产品的单个特性或元件有关。换言之，此时你已经到达了可以描述与特定特性或元件相联系的那个功能层次。将你的最低层功能与产品的单个特性或元件进行比较，如果有些特征或元件没有在功能树上列出，则可能它们与消费者对产品功能的感知没有关系，或者你在功能树上丢失了功能的一整个分枝。

Next, arrange these functions into a "function tree". To start the function tree, select the prime function of the product. This is the main reason that the product exists in the eyes of the customer. Once the prime function has been selected, the other functions are grouped logically and hierarchically under it. The next layer under the prime function should describe "basic function". Basic function relate to prime function in two ways: ①They are essential to the performance of the prime function; ②They are direct cause of that prime function occurring. Keep going through the function tree, asking the question "How is this function achieved?" At every level the functions should be essential to, and direct causes of, the function above. At the bottom of the tree you should end up with a list of functions which cannot easily or logically be subdivided into subsidiary functions. In most function analysis, these lowest order functions relate directly to single features or components of product. In other words, you have reached the level of description of functions which relate to specific features or components. Compare your lowest order functions against the individual features or components of the product. If there are features or components not listed in the function tree, then either they should be irrelevant to the customer's perception of the function of the product or you have missed out an entire branch of functions in your function tree.

拓展阅读

Extend and reading

Tupperware FlatOut! Containers

Tupperware's FlatOut! storage containers were designed for the consumer who needs flexibility in storage and for busy people on the go-students, working professionals and campers. Its accordion design allows FlatOut! to be adjusted to the appropriate capacity for the food to be stored or for space-saving stackable storage. When the container is closed the seals are completely watertight. For cleaning, they can be stacked like plates into a dishwasher and washed flat. In December 2004, the product was awarded a Good Buy Award by the Good Housekeeping Institute and was named by Fortune magazine as one of the best 25 products of the year in their December 2004 issue. It was also recently awarded a "Product of the Year 2005" award by the German Consumer Plastics Institute.

Timberland travel gear

Timberland developed a new concept to address the common dilemma travelers face when deciding how many and which shoes to pack. Often, the shoes they bring with them are not adequate for all the activities they engage in; and what they do bring can consume a significant amount of valuable suitcase space. Timberland travel gear is a revolutionary, modular approach to footwear. By separating the aesthetic(shell)and core function(chassis)of a shoe and making them interchangeable, wearers have access to many more style and function options that can fulfill a range of footwear needs, from hiking to business meetings and dining out to rainy weather. Three pairs of travel gear shells, when combined in

various ways with two chassis and one waterproof hydrosok, are equivalent to twelve pairs of ordinary shoes. Plus, the components pack flat in about half the space of most ordinary shoes.

About the ergonomic

Most products are designed to be used, in some way, by people. When examined in detail, the product-user interface for even the simplest of products is often complex and rarely well understood. Consequently, this aspect of product design often provides a rich source of inspiration on concept design. Task analysis explores the interaction between the product and the person who uses it by observation and analysis and then uses the results to generate new product concepts. It gives the designer first-hand experience of how customers actually use products. Through this, it stimulates concept generation to improve the user interface and paves the way for the subsequent application of ergonomic or anthropometric design methods. Task analysis covers two very important but highly specialized aspects of product development: ergonomic and anthropometry.

The word ergonomic is derived from the Greek word "ergon", meaning work and so is literally the study of work. Early ergonomic studies did, indeed, study people in their working environments but now ergonomic is used much more widely and loosely to refer the interaction between people and built artifacts. Ergonomic is a research topic in its own right and covers aspects of anatomy, physiology and psychology, as well as being applied to design. By far the best way to tackle ergonomic, for most product designers is on a "need to know" basis. If you become involved in a design project which relates to a specific type of interaction with a product, then you start researching current knowledge of the ergonomic of that task. For most design work, however, a sufficient insight into the person-product interaction can be gained by observing people performing the relevant task and, from this, deriving a first-hand understanding of the issues involved.

Anthropometry is the measurement of people. When designing products for people to use, it makes inescapable sense to use measurements of people as basis for product dimensions. The problem with anthropometry is not, generally, finding the data. It is knowing how to use it.

第19课 | 产品开发设计师与设计团队
Lesson 19 | Designers and team for product development

词汇和短语
Words and phrases

creative　创造性的

imaginative　有想象力的

instinctive feel　本能的感觉，直觉

wondrous creature　非凡的人物，精英

virtuoso　大师，名家，学者

perfection　完整性，尽善尽美

what if...?　如果……那将如何呢？

saviour　救世主，救星

team-working　团队工作

prone(to)　倾向

idiosyncrasy　特质，特异性

词汇联想与记忆
Association and memory of words

imaginative	*adj.*	想象的，虚构的
instinct	*n.*	本能
creature	*n.*	人，动物，傀儡，创造物
perfect	*n.*	完成式
	adj.	完美的，理想的，正确的

	vt.	使完美，修改，使熟练
feel	*vt.*	摸，触，感觉，觉得
	vi.	有知觉，有某种感觉
	n.	感觉，觉得，触摸

关键词汇和概念

Key words and concepts

产品开发团队要想高效，就要认清团队成员的强势和弱势，并将他们聚拢在一起互相取长补短；这就是“梦之队”。

Product development team with wish to be effective should recognize the strength and weakness of team members and put them together in ways that complement each other，this is called as the“dream team”.

课文

Text

产品开发设计师与设计团队

Designers and team for product development

我们对新产品设计师都有一种完美的想象。他的创造性和富有想象力使提出的新产品可以改变整个市场。作为视觉技能的典范，他们绘图、建模直至最后产生出的产品具有令人陶醉的视觉形态。凭借对市场的直觉，他们只需有点先见之明地思考一下，就可毫不费力气地生产出消费者梦寐以求的产品。如果这样非同一般的精英跨进你公司办公室的门，产品开发“人员”方面的问题不就立即迎刃而解了吗？你是可以这样想象的！首要和最明显的问题是，这个人是不能单枪匹马地运行这个公司的。需要所有制造、分销、市场拓展和销售等人员与这位新大师一起工作。有非凡才能的人常常会有怨恨别人给他干扰这样一种令人遗憾的习惯，尤其是在他们的创造性自由受到外加的限制时。当富有创造性的设计制造不出来时怎么办呢？产品不能适应现有的销售渠道又怎么办呢？产品过于标新立异，使消费者无法认识到它的价值时又怎么办呢？在寻求尽善尽美时是否就要牺牲开发预算和开发计划时间表呢？即使上面所有这些问题都能克服，更长期些的问题是这位精英设计师在一段时间后一走了之(或被公交车撞倒)时又怎么办呢？

We all have a vision of the ideal designer of new product. Creative and imaginative，they come up with new products which transform entire markets. Paragons of visual skill，they draw，model and eventually produce products with bewitching visual form. And with an instinctive feel for the market they effortlessly produce the product customers would have dreamt of, if only they'd had the foresight to think of it. Should this wondrous creature walk through the door of your company offices, the "people" aspect of product development would be solved in an instant? Or so you might imagine! The first and most obvious problem is that this person cannot run the company single handedly. Manufacturing, distribution, marketing and sales staff will all need to work with the new virtuoso. Exceptionally talented people have an unfortunate habit of resenting interference from others, especially if it involves imposing constraints on their creative freedom. What if the great creation cannot be manufactured? What if they do not fit into existing sales channels? What if the products are so innovative that customers fail to realize their value? Will development budgets and timetables be sacrificed in the search for perfection? And even if all of these problems can be overcome, a longer term issue is what to do when some time later the wonder-designer turns and walks out of the door again(or gets knocked down by a bus)?

鉴于以上种种问题及许多其他原因，光杆司令救世主的经营时代已经过去。取而代之的是团队工作。团队总比任何个人能有更多的时间，他们也有不同的知识和不同的技能，来自于不同的背景。团队的决定也较少有个人决定时的那种古怪的特异性倾向。

For these and many other reasons, the age of the lone business saviour is over. In its place is team-working. Teams have more time than any single individual can have, they have different knowledges and different skills, having come from different backgrounds. A team decision is less prone to the idiosyncrasies of an individual decision.

创建一个高效的产品开发团队，涉及认清潜在团队成员的强势和弱势，并将他们聚拢在一起互相取长补短。梅雷迪恩・贝尔平一直在研究团队的表现。他实验性地创建了由一些特别聪明和富有创造性的人员组成的团队。在他的研究中，他将其称为“阿波罗”团队，暗指这里都是顶(A)级人才。但在专为测试这些团队工作情况而设计的管理训练中，这些团队的表现一直

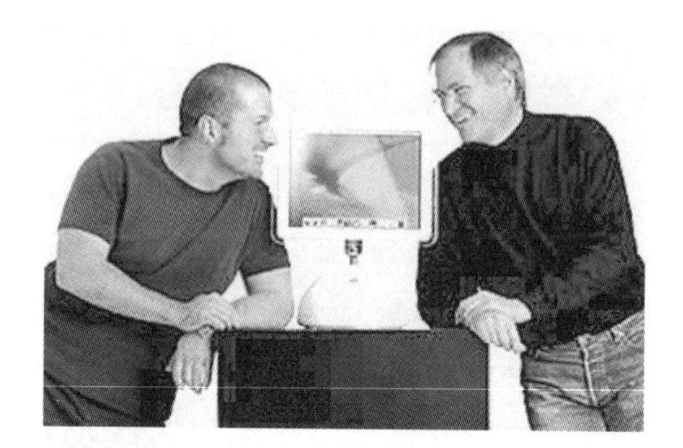

都非常糟糕，比由才能较差的人员组成的团队的表现要差。原因是比起别的团队来，他们难于管理，更倾向于进行破坏性的争论，而且不容易作决定。这使贝尔平进行了关于人们能在一起很好工作的相关特性方面的开创性研究，并确定了“理想团队”的人员混合。

Creating an effective product development team involves recognizing the strength and weakness of potential team members and putting them together in ways that complement each other. Meredith Belbin has worked on teams' performance. He experimentally created teams made up of especially intelligent and creative people. For his research, he labeled these teams "Apollo", cryptically referring to their "A" graded talents. In management exercises designed to test their team-working these teams did consistently badly, performing worse than teams made up of less talented people. The reason was that they were difficult to manage, prone to destructive debates and less able to make decisions than the other teams. This led Belbin to his pioneering work on the characteristics of people that do work well together and he identifies a mix as the "dream team".

拓展阅读

Extend and reading

About Philips Design

Early 1998, Philips Design became an independent unit within the Philips Group, able to provide design services also to clients not only within Philips, but also clients not part of the Philips Group.

Philips Design says that its mission is to create value for customers, shareholders and society as a whole by delivering competitive High Design excellence, to address the quality of life, the ease of and pleasure of owning and using, exceeding customers expectations, promoting technological fusion and meeting manufacturing requirements.

Its vision is to create design focused on personal growth, so that people can live in harmony with each other and with their natural and artificial environments. "Value for people through valuing people."

The design approach is based on the premise that design can never be consistently successful unless it is research-based and people-focused. That's why the company employs so many experts in human science. That's why the company carries out design research projects, often in conjunction with external institutes and partners. That's why the company applies methodologies that go way beyond the boundaries of what many considered to be "design". And that's how the company can answer its clients' needs throughout the business creation process, at every level from conceptual and strategic to product realization and marketing.

Designer Stefan Diez and his some designs

"couch"(see left picture)is made of a cellular textile structure filled with polystyrene balls. The polystyrene balls are added in the country where it's shipped, so it is simply folded for shipping saving on transportation costs.

"tema" was inspired by the bigger proportions of modern service design, since conventional cutlery creations today seem almost invisible next to the big pasta and oversize gourmet plates. Stefan Diez has proven to be a precursor by taking this problem into account. The result is a bigger collection that is reminiscent of the classic form of spade-shaped flatware.

“big bin” is a multi-purpose storage system made up of stackable containers in ABS plastics, suitable as storage boxes for files, toys, laundry, recycling products and transporting of a whole range of items. By combining the various elements like building blocks, a stable shelving system can be quickly created ; the handle grips on the sides of each bin form the interlocking parts for the vertical and horizontal constructions.

When using “genio”, food is taken direct from the kitchen to the table: the stainless steel pots are simply given a porcelain shell, and the food stays warm longer.

Stefan Diez was born in Freising, Munich in 1971. After finishing his carpenters' training he worked in Bombay and Poona, India for one year for a furniture producer and subsequently studied industrial design at the Academy for Fine Arts in Stuttgart. He worked for a short period as assistant designer for Richard Sapper in the US and for Canada two years for Konstantin Grcic before he opened his own studio in Munich in 2002.

第20课 | 创造性原理
Lesson 20 | The principles of creativity

词汇和短语
Words and phrases

principles of creativity　创造性原理

price elasticity　价格弹性

product differentiation　产品的差异化

competitive weapon　竞争武器

competitive advantage　竞争优势

stairway to creativity　创造性阶梯

common sense　常识

first insight　第一洞察力

incubation　深思熟虑

illumination　阐明

verification　查证确认

to the full　充分

parametric analysis　参数分析法

problem abstraction　问题提取

词汇联想与记忆
Association and memory of words

intelligence　*n.*　智力，聪明，智慧

imagination　*n.*　想象，想象力，空想

leave	*n.*	许可，同意，请假，休假
	vt.	离开，动身，遗忘，剩下，委托
leave... for(...)		离开……去(……)
		把……留给(……)
	vi.	出发，动身，生叶
fit	*n.*	适合，痉挛，突然发作
	adj.	合适的，恰当的，健康的
	vt.	适合，安装，使适应，使合格
	vi.	适合，符合
fit in with		与……配合，与……一致，适应

关键词汇和概念

Key words and concepts

创造性是人类最神秘的能力之一；它需要智慧和想象，而不仅是机械的技能。只要足够努力，逐渐掌握其规律，每个人都可以是有创造力的。

Creativity is one of the most mysterious of human abilities. Creativity requires intelligence and imagination, not merely mechanical skill. Everyone can be creative if they try hard and understand principles of creativity gradually.

课文

Text

创造性原理

The principles of creativity

创造性是人类最神秘的能力之一。我们真的更聪明些吗？是否有任何方法可以刺激创造性，或者创造性是与生俱来的吗？心理学家的回答是：我们是更为聪明，你能刺激创造性，而创造性在很大程度上并非与生俱来。因此只要足够努力，每个人都可以是有创造力的。

Creativity is one of the most mysterious of human abilities. Are we really any the wiser? Are there ways of stimulating creativity or is it something you are born with? The psychologist's answers are that yes we are wiser, yes you can stimulate creativity and no, people are not to any significant extent born with in-built creativity. So, everyone can be creative if they try hard enough.

创造性是设计的核心，且贯穿于整个设计过程的各个阶段。在当今大部分市场上，竞争之激烈给价格弹性几乎没有留下空间。因此，纯粹以价格为基础的竞争已极其有限，产品的差异化已成为产品竞争的主要武器。通过产品差异化来保证竞争优势，就意味着在你的产品和你的竞争者的产品之间创造不同。

Creative is at the heart of design, at all stages throughout the design process. In most markets today there is sufficient competition to leave little room for price elasticity. Competition purely on the basis of price is, consequently, severely limited and therefore, product differentiation remains the main competitive weapon.Securing a competitive advantage through product differentiation means creating differences between your products and those of your competitors.

创造性思维是创新的摇篮
Creative thinking is the cradle of the innovation

对创造性的常识性理解的一种结构性表达方法，可以用创造性阶梯来进行描述。创造性阶梯可以分为五个台阶：第一洞察力、准备、深思熟虑、阐明和验证。

A structured way of presenting a common sense understand of creativity can be described as the stairway to creativity. The stairway to creativity can be of 5 steps, which are the first insight, preparation, incubation, illumination and verification.

第一洞察力是在你头脑中对某些创造性发现的需求进行架构的方法。在一个范围内架构出问题，集中精力于该问题上直至问题解决。

The first insight is the way you frame in your mind the need for some creative discovery. Frame the problem in one area and focus their minds on that problem until it is solved.

设计问题一般是复杂的，会有很多个目标，很多限制，也就有更大量的可能解决方案。设计新产品时，你会试图满足很大范围的消费者需求，充分开拓营销、市场和分销渠道的能力，适应现有制造设备和供应商，最终为公司创造利润。考虑所有这些因素来确定一个设计问题，需要作很多准备。准备工作包括：吸收所有有关的事实和概念，为实现创造性突破加油，探索、扩展及确定问题，将现有方案的可能性全部列出来。

Design problems are usually complex, in as much as they have several goals, many constraints and an even greater number of possible solutions. In designing a new product, you will be trying to satisfy the needs of wide range of customers, exploit to the full abilities of sales, marketing and distribution channels, fit in with existing manufacturing facilities and suppliers and end up making a profit for the company. Defining a design problem to take account of all this takes a lot of preparation.

Preparation involves absorbing all the relevant facts and ideas to fuel the creative breakthrough, exploring, expanding and defining a problem and exhausting the possibilities of existing solutions.

深思熟虑，就是在你脑子里对实情进行消化、吸收和分类。为了孕育一个想法，你需要让问题先驻留在你脑子里。身心放松并让你的思维漫游，这样，更多变化多端的思维就会接踵而来，而这中间的某些想法就可能打破常规或形成非正统的联系从而打破创造性的壁垒。

Incubation involves digesting, assimilating and sorting all facts in your mind. To incubate an idea, you need to let the idea "settle" in your mind. By relax and letting your mind wander, more diverse thoughts come to mind and some of these diverse thoughts just might make an unusual or unorthodox connection that breaks through the creative wall.

解决问题的阐明阶段必须要积极搜寻，并需要一些专门技术来加快这个过程。

The illumination part of problem solving must be actively sought and therefore techniques are needed to accelerate the process.

在探索、扩展和确定问题时有两种不同的方法：①参数分析法，它探索问题的定量、定性和完备的测量。参数分析法往往起关键作用。②问题提取，它简化问题以逐步地提炼出概念。假设问题如最初陈述时那样，则问题提取就会问，为什么一开头就想解决这个问题呢。这是个找到问题根源的好方法。

There are two different methods for exploring, expanding and defining problems: ①Parametric analysis, which explores the quantitative, qualitative and categorical measurement of the problem. Parametric analysis usually plays a key role. ②Problem abstraction, which tries to reduce the problem to progressively abstract concepts. Given a problem as initially stated, problem abstraction asks why you want to solve that problem in the first place. This is a good way of trying to get the root of a problem.

拓展阅读

Extend and reading

Making technology warm and fuzzy

High-tech consumers are forming an unlikely alliance with craftspeople in an effort to buck the mainstream when it comes to aesthetically pleasing electronics accessories.

Geeks crave more individuality in their accoutrements than Apple and others can offer with a handful of bright colors. So a raft of handmade items has entered the online marketplace, featuring hand-stitched felt with sewn-on hearts and stars, fabric festooned with yellow birdies and bumblebees, and crocheted cozies.

Take Vanessa Brady, who after purchasing her iPod last year searched for a case that would protect the gadget while showing off her stylish sensibility. Unimpressed with mass-manufactured cases on the market, she gave up on the mainstream and tried her hand at making iPod holders that she—and eventually others—preferred, both for their aesthetics and their protective qualities.

Brady, 27, is creator and owner of Gerbera Designs. She lives in El Paso, Texas, and sells her iPod cases for $30 to $35 on her own site, as well as places like PixelgirlShop.

The trend proves that crafty and techie are not mutually exclusive personality traits. Since most crafters like Brady use their own products, consumers can be sure that while the product is handmade, it will also be protective of their iPods or laptops.

From Seattle to Brooklyn, crafters are creating reasonably priced items like carrying cases and protective gears for electronics. Because they're made by hand, the products are often in limited supply, and, unlike their mass-produced competitors, no two are exactly the same.

This isn't part of a new crafting revolution, but rather crafters adapting to whatever new technology comes along, said Shoshana Berger, editor in chief of *ReadyMade*, a do-it-yourself magazine.

"I think when the teapot came around, people started making tea cozies, and now people are making iPod cozies", Berger said.

Whether crafty themselves or not, consumers seem to appreciate pairing technology with handmade accoutrements. People interested in buying these products are gadget hounds in their 20s and 30s, Berger said, some of whom are interested in buying crafty objects because gadget life cycles are so short.

"We have planned obsolescence built into these products, and your iPod is outdated in a year or two. Just based on our taking the temperature of our readers, people are a little bit alarmed at that and like the idea of having a handmade element in their lives", she said.

According to a survey by the Craft Organization Development Association, as of 2000, estimates showed over 126000 people in the United States earned some or all of their incomes from selling handmade goods.

About 70 percent of professional crafters sell goods on their own websites, and the number is increasing, said Ann Barber, membership director for the National Craft Association.

Such products also give consumers a way to personalize gadgets that are often starkly silver or black, said Janice Headley, 28, who runs Copacetique, where she sells handmade products made by herself and others. Headley, who lives in Seattle, sells $34 felt CD wallets adorned with things like record players and zoo animals.

The wallet and other products she uses—a crocheted cell-phone holder, a pink-and-white handmade laptop bag—are ways of personalizing technology, something that appeals to both consumers and crafters.

"People want variety, people want things to be personalized. I think since the electronics companies aren't really providing that, people are taking it upon themselves. If you can't buy a laptop in pink, you make a laptop case in pink," Headley said.

Rob Kalin is a co-founder of the online craft market Etsy, which opened in June and has 1700 crafters registered to sell goods on the site. "It's only until recently that the fact that things are handmade made them a niche market. That used to be the only market," he said.

In the end, though, no matter how high-or low-tech the item, maybe it just comes down to its geek-chic factor. Patricia Valdez, 27, said she gets plenty of comments on her robot-patterned Gerbera Designs' iPod case.

"I think it's really cool that people notice it," she said, "It's unique and different."

第6单元
Unit 6

第21课 | 通用设计
Lesson 21 | Universal design

词汇和短语
Words and phrases

universal design　通用设计

people with disability disables　残障者

indispensable　必不可少的

independence　独立自主

assistive design　帮助设计

accessible design　接近性设计，亲近设计

adoptive design　适应性设计

barrier-free design，design without barriers　无障碍设计

hospitability design　关怀设计

transgenerational design　跨代设计

life-span design　全寿命设计

integrative design　整合设计

non-discrimination　无歧视

equal opportunity　机会均等

personal empowerment　个人权利

democratic value　民主价值

user-centeredness　用户中心论

词汇联想与记忆

Association and memory of words

universal	*adj.*	普遍的，全体的，通用的；宇宙的，世界的
disability	*n.*	无力，无能，残疾；[律]无能力，无资格
discrimination	*n.*	辨别，区别，识别力；辨别力，歧视
equal	*adj.*	相等的，均等的，不相上下的；
	n.	对手，匹敌，同辈
empower	*v.*	授权于，使能够
democratic	*adj.*	民主的，民主主义的；民主政体的，平民的

关键词汇和概念

Key words and concepts

通用设计，简单地说就是为所有人的设计；它需要面对所有人(不论其身体状况、年龄、残障程度如何)，并设计出人人都能方便使用的产品和环境。

Universal design，in one word，is design for all people，it should face to all people(regardless the physical condition，age，disability)and design the products and environments being used by people easily.

课文

Text

通用设计

Universal design

在通用设计之前，已经有好几种为残障人服务的设计方法了。

Before universal design，there are several design methods for people with disability disables.

帮助设计创造专门的产品来增加、维持和改善伤残个体的功能性能力。帮助设备包括许多移动辅助设备(例如轮椅，带轮拐杖，步行杖和步行扶车等)和洗手间帮助设备(例如高度升高的马桶和小便器及扶手栏杆等)。这些对残障人必不可少的产品帮助他们达到独立。

Assistive design creates specialized products used to increase, maintain or improve the functional capabilities of individual with disabilities. Assistive devices include mobility aids(such as wheelchair, walking sticks, crutches and walker)and toileting aids(such as raised seats and grab bars). Indispensable for helping people with disabilities achieve independence.

接近性设计则为满足伤残人的特殊需要而制作特殊的装置。例如较宽的门框，较高的洗脸盆，较低的喷泉式饮水器和投币电话等，它们可在已建成的环境中提供可接近性。它常常通过“复制”现有产品(某些部分常经过改造)来体现对残障人的关注。

Accessible design produces specialized devices that meet the unique needs of people with disabilities. Wider door's widths, higher lavatories, lower drinking fountains and pay phones are examples of designs that provide accessibility into built environment. It often "copies" the existing product(some part of the product often has modified)to impersonate the attentions given to the people with disabilities.

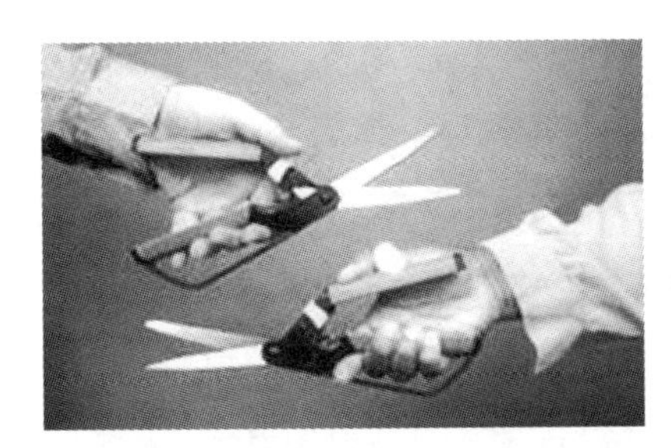
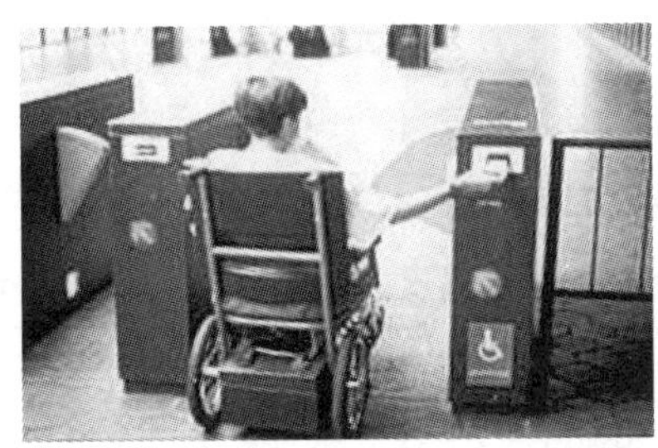
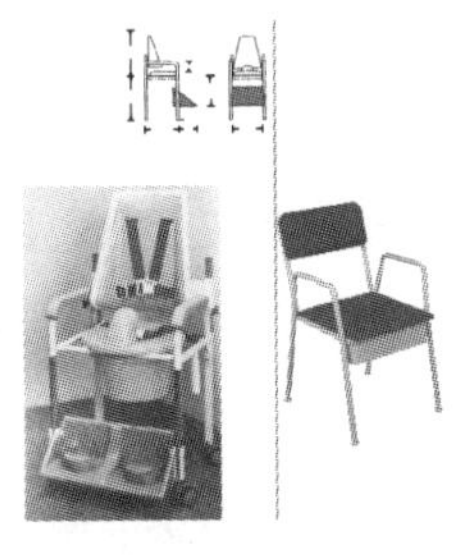

接近性设计则为满足伤残人的特殊需要而制作特殊的装置
Accessible design produces specialized devices that meet the unique needs of people with disabilities

适应性设计则使普通产品能让能力有缺陷的人群(老人、儿童、伤残人)使用。例如杠杆式门把手上加装的滑条，专用的餐具和炊具等。

Adaptive design makes common products usable by people with disabilities(old people, children and people with disabilities). Like the slip on lever handle, modified silverware or a stove etc.

帮助设计、接近性设计和适应性设计的意义比较接近，有时也用关怀设计和无障碍设计这样的术语。其实它们之间是有微小差别的。

Assistive design, accessible design and adoptive design are somewhat similar in their meanings, sometimes the term hospitability design and design without barriers are also used. However, there are minim differences in between of these designs.

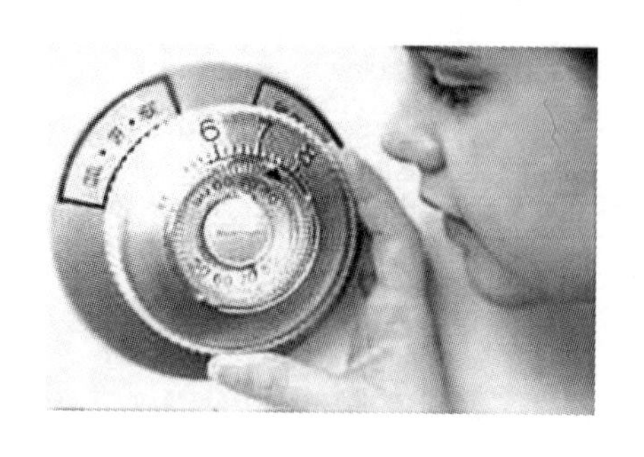

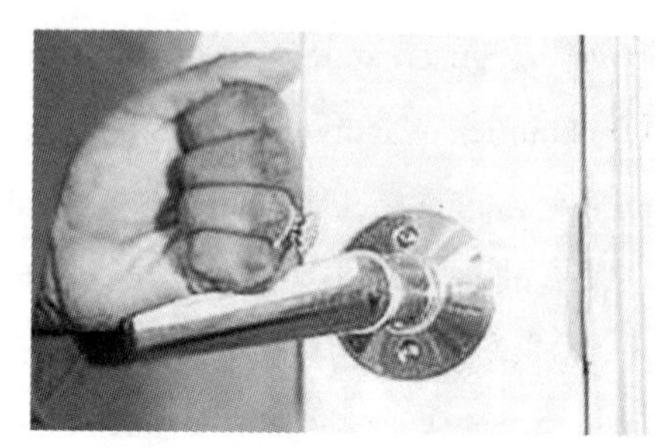

通用设计，简单地说就是为所有人的设计；它需要面对所有人(不论其身体状况，年龄、残障程度如何)，并设计出人人都能方便使用的产品和环境。

Universal design，in one word，is design for all people，it should face to all people(regardless the physical condition，age，disability)and design the products and environments being used by people easily.

通用设计与上面提及的几种设计有很大的不同。通用设计与其他设计方法的区别在于它重视社会内涵。它在社会层面上立足于无歧视、机会均等和个人权利的民主价值。事实上它是通过多样性来体现个性化，在同一个系统中为不同的使用者做不同的设计。然而，这只是个理想而已。事实上，通用设计从未规定过最终状态，因为实在没有一样东西是真正万能和通用的，而产品的演进也永无止境。通用设计高度的社会目标也几乎不可达到，亦即，为“所有人”设计“所有的事”的目标几乎是不可能达到的。因此，其目的只是通过“通用设计”朝此目标努力而已。因此从这个观念上说，通用设计与上述几种设计存在着千丝万缕的联系；在实践中更无法与那些为特定人群所进行的设计隔裂开来。

While universal design is very different from the designs mentioned above. What separates universal design from other design approaches is its focus on social inclusion. The social aspect of universal design is grounded in democratic value of non-discrimination，equal opportunity and personal empowerment. Universal design is，in fact，about individualization through diversity，different design for different users within the same system. This is，however，only an ideal. In fact，universal design does not prescribe to a final state，since nothing is truly universal nor is product evolution even finished. The high social objectives of universal design are almost unattainable，that is，it is nearly impossible to design all things for all people. The goal，therefore，is to approach these objectives through “universal designing”. Therefore，universal design has countless ties with other designs mentioned above in its concept，it is even more difficult to separate it from other designs designing for the specified group of people in practice.

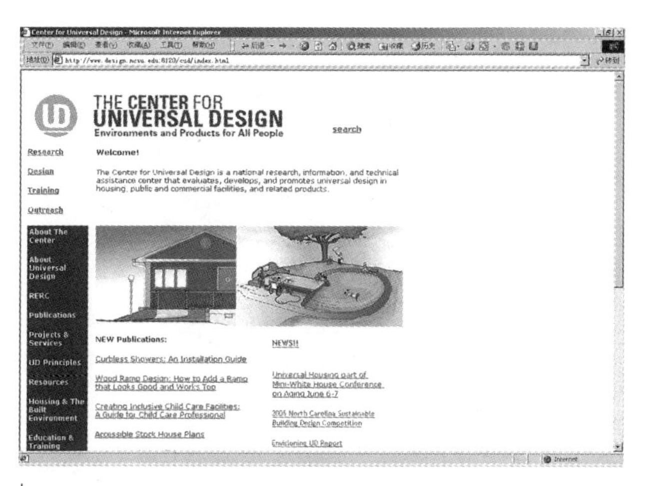

通用设计网站之一 One of universal design websites.

通用设计的“七项原则”

Seven principles of universal design

(1) 公平的使用性；

(1) Equitable use;

(2) 使用的灵活性；

(2) Flexibility in use;

(3) 简单直观性；

(3) Simple and intuitive;

(4) 信息明显性；

(4) Perceptible information;

(5) 错误容忍性；

(5) Tolerance for error;

(6) 省力的使用性；

(6) Low physical effort;

(7) 接近和使用的尺寸和空间的适当性。

(7) Size and space for approach and use.

我们每个人都会是“残障人”。社会上人人都是平等的，我们都需要关爱。

Each of us sometime will be person with disability. Every one in the society is equal, we need love and care.

通用设计举例：浴室。

Example: a bathroom.

拓展阅读
Extend and reading

Self-watering flowerpot

This self-watering flowerpot assures plant lovers that their plants won't die when they are on vacation or busy with their hectic lives. The design is aesthetic as well as functional. A wick is placed in the holes of a ceramic flowerpot. The plant is placed in the pot, which is then placed on top of a glass container filled with water. The wick dangles into the container and the plant can then take up the necessary amount of water through the wick. The glass container holds enough water for a week, on average. The design gives the user simple, straightforward feedback. One glance at the water lever will tell you if more water is needed.

MINI_motion watch

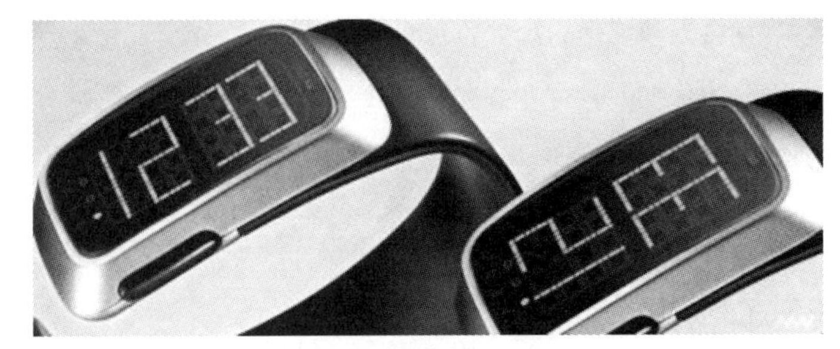

The MINI_motion watch features a digital display that changes from horizontal to vertical, ergonomically facilitating easy reading when the wrist is in-line with the user's vision whether riding a bike, walking or driving. The watchband is adaptable to different wrist sizes, easy and quick to put on and off, and comfortable, thanks to an elastomer soft band over a spring steel core. The watch is sold through high design outlets such as Barneys, Flight 001, MOMA and SFMOMA stores, as well as an array of small retail stores and catalogues, and is one of the best selling items in the NY MOMA catalogue. The product sold-out quickly and is currently in its second production run. Based on the product's success, MINI plans to release additional MINI_motion watch styles and sizes in 2005.

K2 T1 boot with Boa liner

The purpose of the project was to reinvent K2 Snowboard's top of the line T1 boot, integrating new technological and advances styling. Snowboard boots traditionally have two lacing systems, one for the internal liner and another for the shell(or exterior). The T1 has the Boa Access liner, which allows the rider to fine tune the tension without unlacing and removing the shell. To tighten, the Boa dial on the upper cuff is turned while popped in. To instantly release the tension on the cable system, the rider just pops the Boa dial out. The 2004~2005 K2 boot line, of which the flagship model is the T1, has increased sales volume by 25 percent over the previous season.

The taste of design

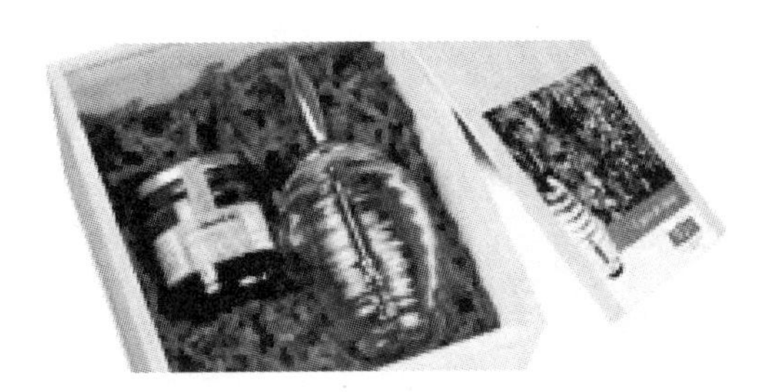

Excellence and passion: thanks to this dual concept Alessi and the Slow Food Foundation for Biodiversity Onlus singled each other out and united to promote an important initiative, featuring new and original combinations of gourmand pleasures and design creativity.

Ten food products—chosen among over 270—protected by Italian and international Slow Food Foundation Presidia. Saffron, salt, lentils, rice, honey, coffee, vanilla, almonds, figs, cornmeal biscuits, all originating from local micro-productions, are offered to the public with an Alessi item suited to hold and serve them with a style fit for that "added" value. Diligent simplicity and elegant naturalness also accompany the packaging that includes a brochure with the history, purpose and logistics of the biodiversity projects.

The Slow Food Foundation for Biodiversity, always attentive and selective when choosing its partners, follows a common path with Alessi that is the indivisibility between the passion to "do" and the culture of "doing it well".

With the distribution in its retail chain at Christmas time, Alessi teams up to fully and actively support these first small ten food manufacturers who, directly and without any economic intermediaries, provide their products and who, thanks to this operation, can count on an important and well qualified market opening with an attentive public, appreciative of flavour and quality.

第22课 | 生态设计：概念和原理

Lesson 22 | Ecological design：concept and principles

词汇和短语

Words and phrases

ecological　生态的，生态学的

environment　环境

destructive　破坏(性)的

discipline　训练；学科

harmonize　(with...)协调

landscape　景观，风景

aesthetics　美学，审美学

prudence　审慎；节俭

intimate　亲密的，隐私的；熟友

survival　生存，幸存；幸存者

symbiosis　共生(现象)，互利合作关系

mass　块；群众；大量，大规模的

dissolve(into...)　解散；溶解

luxury　豪华的，奢侈的，华贵的

ethic　伦理，道德规范

responsibility　责任(性)，职责

moral　道德；精神上的

conflict　矛盾，冲突，抵触

imperfection　不完整性

deviation　背离，偏差

insight　洞察力，见识

disengage (from...)　脱离……

dissever　割裂，分开

diverse　多变化的，不同的

sustainable　可持续(发展)的

词汇联想与记忆
Association and memory of words

ecology　*n.*　生态学，[社会] 环境适应学，均衡系统

structive　*adj.*　结构的，建筑的

response　*n.*　回答，响应，反应

sustainable　*adj.*　可支撑的，可撑住的，可维持的，可持续的

diversification　*n.*　变化，多样化

关键词汇和概念
Key words and concepts

生态设计是通过设计与生态过程整体化考虑，尽量使其对环境的破坏性影响减到最小的一种设计。简单地说，生态设计是对自然过程的有效适应及结合。

Ecological design is a design that minimizes environmentally destructive impacts by integrating itself with living processes. In short, it is an effective fitness to and combination with natural processes.

课文
Text

生态设计：概念和原理
Ecological design: concept and principles

任何通过设计与生态过程整体化考虑、尽量使其对环境的破坏性影响减到最小的设计形式，都称为生态设计。生态设计不是某个职业或学科所特有的，它是一种与自然相互作用和相互协调的方法，其涉及范围非常广，包括建筑和景观、工业产品、工业流程等。生态设计帮助我们重新审视设计及人们的日常生活方式和行为。简单地说，生态设计是对自然过程的有效适

应及结合。

Any form of design that minimizes environmentally destructive impacts by integrating itself with living processes is called ecological design. Ecological design is not specially owned by a profession or a discipline，it is an approach to harmonize and interact with nature and it is fargoing，including architecture and landscapes，industrial products，industrial processes and so on. Ecological design helps us to review our designs and people's daily life styles and behaviors. In short，it is an effective fitness to and combination with natural processes.

生态设计反映了人类的一个新理想，一种新美学观和价值观：人与自然的合作与友爱的关系。

The ecological design reflects the new ideal of human being，and new aesthetics and values：to be friendly and to cooperate with nature.

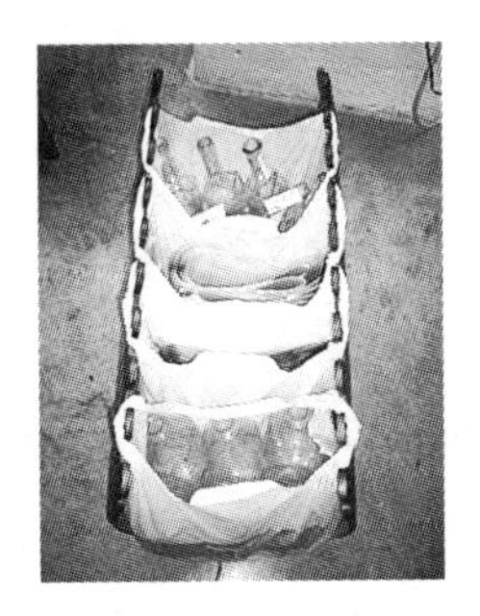

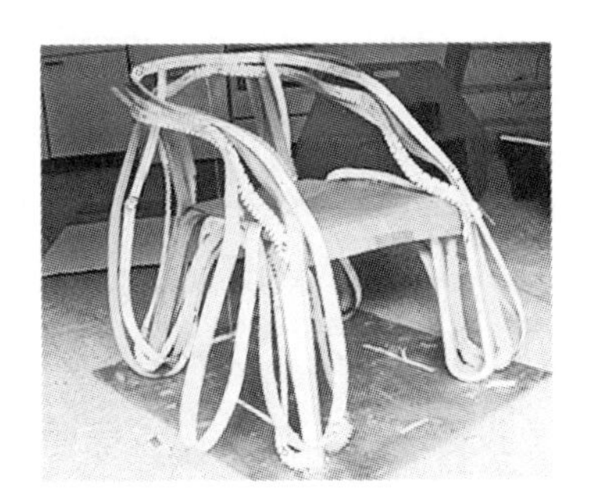

下面讨论其四条基本原理：地方性设计，节俭的设计，与自然一致的设计，以及使自然可视化。

Four basic principles are discussed here：design with place，design with prudence，design with nature and make nature visible.

地方性设计。生态设计始于对一个地方的熟识。设计师首先应该考虑的问题是我们在什么地方？自然允许我们做什么？自然又能帮助我们什么？我们还应尽量使用当地的自然材料。

Design with place. Ecological design begins with the intimate knowledge of a place. Designers should first consider where we are，what nature allows us to do and what help nature can provide us? We should try our best to use local natural materials.

节俭的设计，即保护与节约自然资本。因此对于自然生态系统的物流和能流，生态设计强调保护、减量、再利用和再生。

Design with prudence，i.e. to protect and save natural capitals. Therefore ecological design emphasizes protection, reduce, reuse and recycle in the material flow and energy flow in natural ecological system.

与自然一致的设计，即让自然起作用。自然生态系统为维持人类生存和满足其需要提供了各种条件和过程，此即生态系统的服务。这一原理强调人类与自然过程的共生和互利合作关系；通过与生命所遵循的过程和模式的协调，我们可能显著减少设计对生态的影响。

Design with nature, i.e. let nature do work. Natural ecological system provides all kinds of conditions and processes to maintain human survival and satisfy their demands, this is so called services of ecological system. This principle emphasizes the symbiosis between human being and nature processes, and it is possible for us to reduce obviously the ecological impact arising from design through cooperation with processes and patterns followed by life.

使自然可视化。生态设计回应了人们对土地和土地上的生物之依恋关系，并唤醒人们对自然的关怀。自然循环和过程的可视化把设计环境带回生活。

Make nature visible. Ecological design responds to the attaching relationship of human beings with land and biology on land, and wakes up human beings to care of nature. Making natural cycles and processes visible brings the designed environment back to life.

此外，生态设计还强调人人都是设计师和参与者。传统设计强调设计师的个人创造，认为设计是一个纯粹的、高雅的艺术过程。而生态设计既是人与自然的合作，也是人与人合作的过程。因此生态设计强调人人皆为设计师，因为每个人都在不断地对其生活和未来作决策。所以，从本质上讲，生态设计包含在每个人的一切日常行为之中。对专业设计人员来说，这意味着自己的设计必须走向大众，走向社会，融大众的知识于设计之中。同时，使自己的生态设计理念和目标为大众所接受。

Besides, ecological design also emphasizes that everyone should be a participant-designer. Traditional design emphasizes individual creation of designers and is considered as a pure, elegant artistic process. Ecological design is a process of cooperation between human and nature, and between people and people. Therefore ecological design emphasizes that everybody is designer as everybody has been constantly making decisions for his life and future. Therefore essentially, ecological design is included in everyone's day to day behaviors. This means to professional designers that design should be taken to the masses and to the society and dissolve public acknowledge into designs. At the same time, they should make their own ideas and aims of ecological designs acceptable by the masses.

生态设计不是一种奢侈，而是必需；生态设计是一个过程，而不是由专业人员提供的一种产品。生态设计更是一种伦理；它反映了设计者对自然和社会的责任，是设计师的崇高职业道德的体现。

Ecological design is not luxury but necessary, it is a process and not a product given by professionals. Ecological design is an ethic, it reflects the responsibility of designers for nature and society and grand professional moral of designers.

生态设计应该是经济的，也必须是美的。生态和经济本质上是同一的，生态学就是自然的经济学。两者之所以会有当今的矛盾，原因在于我们对经济的理解的不完全性和价值偏差。生态设计原则给我们生活方式的再设计提供了洞察力、灵感和指导。它照亮了一条道路，使我们能脱离共同的消费文化，创立更加充实的生活方式。

Ecological design must be economic and beautiful. Essentially ecological design and economy is identical. Ecology just is nature's economy. The currently existed conflict between ecology and economy is due to imperfection and value deviation of our understanding about the economy. Principles of ecological design give the insight, inspiration, and guidance for a redesign of our way of life. It illuminates a way to disengage from corporate consumer culture, and create a more fulfilling lifestyle.

与传统设计相比，生态设计在对待许多设计问题上有其自身的特点。但是，生态设计应该是传统设计方法的进化和延续，而非突变和割裂。缺乏文化内涵和美感的生态设计是根本不能被社会接受的。生态设计应该、也必须是美的。

Comparing to traditional design, ecological design owns its specialities towards many design aspects. However ecological design should be an evolution and extension of traditional design approach, but not as any break or dissevering. Ecological design without any cultural meaning or aesthetic will not be acceptable by the society at all. Ecological design should and must be beautiful.

我们相信，自然世界和人类设计的世界可以达到重新整合。应予发扬的生态设计原则是：人类和自然应该以健康、支持、多样和可持续发展的方式共存。

We believe that the natural world and the humanly designed world can be reintegrated. The ecological design principles being promoted are: humanity and nature should co-exist in a healthy, supportive, diverse and sustainable condition.

拓展阅读

Extend and reading

Customer demand and new product development

Customer demand, competitor actions and more stringent safety and environmental legislative changes compel enterprises to develop new products from time to time. The managerial tactic of consciously shortening the market life of the product, through rapid new introduction is a strategic weapon against slower-moving competitors.

New product development falls into three groups: one is simply to improve or modify the existing product, other is an integrated development, and the last is technical innovative development.

Developing new products is both important and risky. But why is it so risky? What makes so many products fail? In new product development terms the uncertainties are highest at the start. You don't know what the product is going to look like, how it will be made, what it will cost or what customers will think of it. In conclusion, developing new products is a multi-factorial problem, success or failure is determined by many factors(e.g: attraction to customers, acceptability to retailers, engineering feasibility, product durability and reliability).

To sum up, a number of factors in three broad areas made the most important differences between success and failure:

(1) Market orientation: the biggest, probably most obvious factor determining commercial success was market differentiation and customer value. Products which were seen by customers as being substantially better than competing and being better in ways which were highly valued had 5.3 times the success rate of those that were only marginally different. Other factors contributing to market orientation were found to be ability to market the product early and to having a highly effective product launch.

(2) Early feasibility assessment and specification: two factors are important here. Products which underwent a thorough and stringent assessment prior to development were 2.4 times as likely to succeed as those that had not. Products which were sharply and well defined in design specification prior to development were 3.3 times as likely to be successful as those that were not.

(3) Quality of the new product development process and the team that does the work: where technical activities are consistently carried out to a high quality, products have, in general, a 2.5 times greater success rate. Specifically, where the company's technical skill were well matched to the activities needed to develop the new product, the chances of success were 2.8 times greater. Where the company's sales and marketing skills were well matched to the new product the chances of success were 2.3 times greater. Where there was a high degree of "working harmony" between the technical and marketing staff within a company, the chances of product success were 2.7 times

greater than where there was “severe disharmony”.

To make a new product development success, usually a risk management funnel is used to manage the risk. A risk management funnel is a way of thinking about new product development which shows how risk and uncertainty change as new products develop, it represents the decision-making about a new product during its development.

第23课 | 飞利浦设计公司的高端设计方法
Lesson 23 | Philips Design's high design process

词汇和短语
Words and phrases

seamlessly　无缝的，紧密无间的

design skill　设计技巧，设计技能

psychologist　心理学家

sociologist　社会学家

anthropologist　人类学家

era　时代，纪元

mission　使命，任务

client　客户，委托人

shareholder　股东

in harmony with...　与……协调、和谐

premise　前提，假定

in conjunction with...　与……联合

methodology　方法学，方法论

enable(enebling)　使能够，授权

finalization　定案，定稿

词汇联想与记忆
Association and memory of words

integration　*n.*　综合

integration of...into...		……与……的整合
integrate	*v.*	结合
	vt.	使成整体，使一体化
integrate...into...		把……综合为……
psychologist	*n.*	心理学者，心理学家
sociologist	*n.*	社会学者，社会学家
anthropologist	*n.*	人类学者，人类学家
human-focused	*adj.*	关注人的，以人为本的

关键词汇和概念
Key words and concepts

与传统的设计方法比较，飞利浦设计公司的高端设计方法是以人为本的，多学科融合的及以研究为基础的设计方法，它使得设计过程紧密地整合到商业创造过程中去。

Comparing with traditional design, Philips Design's High Design is a human-focused, multi-disciplinary and research-based design approach which allows the seamless integration of design process into business creation process.

课文
Text

飞利浦设计公司的高端设计方法
Philips Design's High Design Process

从1991年被任命为(飞利浦设计的)负责人起，斯德芬纳·玛扎诺相信，对于创造“适合于人们日常生活相关的有内涵的解决方案”单靠设计技能已经不够了。他执行了一种以研究为基础、并与强烈的以人为本相结合的策略，称之为“高端设计方法”。这个设计过程完全融合进商业过程中，并将诸如趋势分析师、心理学家、社会学家和文化人类学家与设计相关的技能包括进去。玛扎诺持续地以这样一种信心来推动自己：解决方案不仅是由于技术上的可能性创造出来的，而且来源于人们希望以他们自己喜欢的方式来改善他们的生活质量。

From his appointment in 1991 as director, Stefano Marzano believed that design skills alone were no longer enough to create “relevant, meaningful solutions that fit into people's everyday lives”. He implemented a research-based strategy with a strong human focus, called “High Design”.

The process is fully integrated into the business process and includes input from design related skills, such as trend analysts, psychologists, sociologists and cultural anthropologists. Marzano continues to be driven by the belief that solutions should not be created just because they are technologically possible, but because people want them because they improve their quality of life in the way they themselves would like.

飞利浦眼中的过去、现在和将来
Past, present and future from Philip's eyes.

1991年斯德芬纳·玛扎诺以发展和导入新的观点和任务为起点，成功地开创了飞利浦的设计新纪元。这个新观点和新任务着重于通过他们称之为的“高端设计方法”为人们创造价值。高端设计方法是一种以人为本、多学科融合和以研究为基础的设计方法，它使得设计过程紧密地整合到商业创造过程中。

In 1991, Stefano Marzano succeeded a new era of design at Philips started with the development and introduction of a new vision and mission. This new vision and mission focuses on creating value for people through what they call “High Design”. High design is a human-focused, multi-Disciplinary and research-based design approach which allows the seamless integration of the Design Process into the business creation process.

1998年初，飞利浦设计成为飞利浦集团旗下的一个独立单位，不仅能为飞利浦集团内部的客户提供设计服务，也可以为飞利浦集团外的客户提供设计服务。

Early 1998, Philips Design became an independent unit within the Philips Group, able to provide design services also to clients not only within Philips, but also clients not part of the Philips Group.

飞利浦设计认为该公司的任务是：通过提供有竞争力的、卓越的高端设计，为消费者、公司股东及整个社会创造价值；超越消费者的预期，促进技术的融合，满足制造的要求，致力于提升生活质量以及拥有和使用的方便和愉悦。它的观点是，创造关注个人成长的设计，以使人们能够彼此并与自然和人造环境和谐地生活。“通过评价人来达到为人创造价值。”

Philips Design says that its mission is to create value for our customers, our shareholders and society as a whole by delivering competitive High Design excellence, to address the quality of life, the ease of and pleasure of owning and use, exceeding customers expectations, promoting technological fusion and meeting manufacturing requirements. Its vision is to create design focused on personal growth, so that people can live in harmony with each other and with their natural and artificial environment. “Value for people through valuing people”.

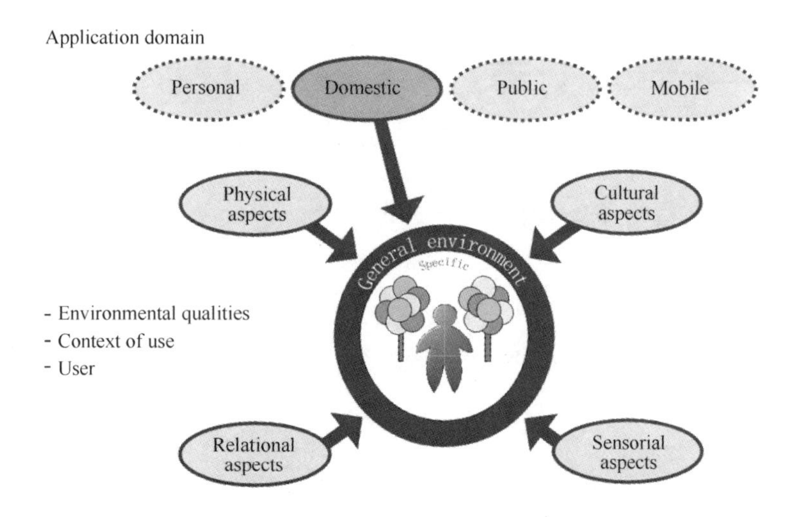

这种设计方法基于这样一个前提：除非设计以研究为基础并以人为中心，否则设计决不可能次次成功。这就是为什么该公司雇佣了这么多人类科学方面的专家的原因。这就是为什么该公司常常与外面的研究单位和合作伙伴联合一起进行许多设计研究项目的原因。这就是为什么该公司要采用一些被很多人认为是超出“设计”之外的方法的原因。这也就是为什么公司能够通过商业创造过程在各个层面上，从概念和战略直至实现产品和市场层面上，回应客户的需求。

The design approach is based on the premise that design can never be consistently successful unless it is research-based and people-focused. That's why the company employs so many experts in human sciences. That's why the company carries out design research projects, often in conjunction with external institutes and partners. That's why the company apply methodologies that go way beyond the boundaries of what many consider to be "design". And that's how the company can answer its clients' needs throughout the business creation process, at every level from conceptual and strategic to product realization and marketing.

PHILIPS

the multi-disciplinary approach by integrating design & business capabilities

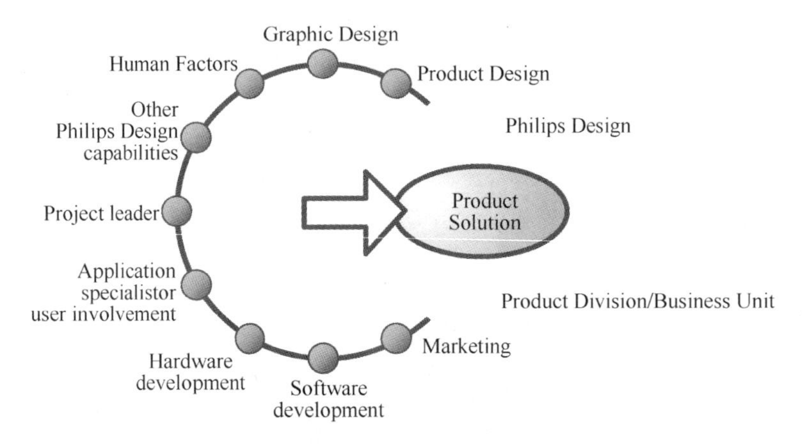

飞利浦设计公司的多学科方法
The multi-disciplinary approach of Philips Design.

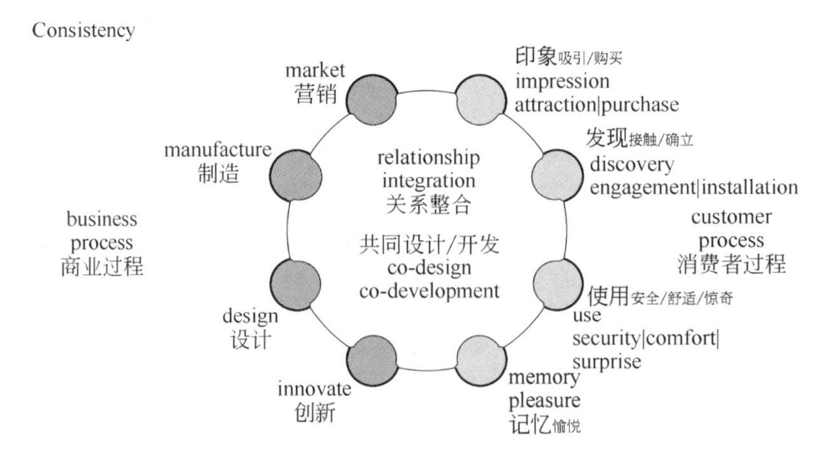

将商业过程与消费者整合进来的共同设计和共同开发
Co-design and co-development after business process and customer process are integrated.

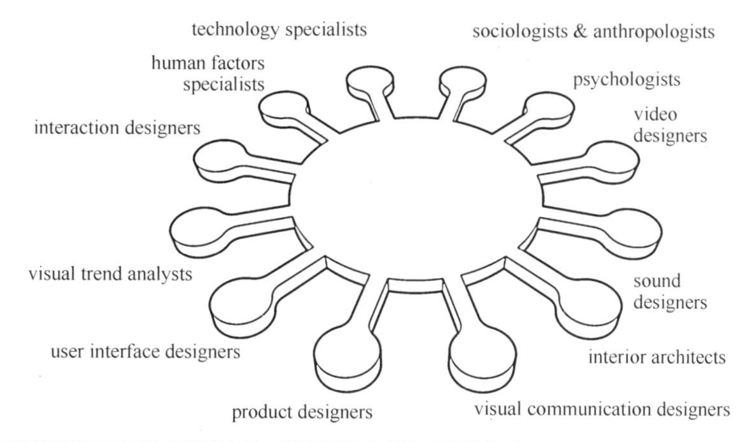

飞利浦设计公司的多学科方法：设计团队人员的多学科组成
The multi-disciplinary approach of Philips Design：design team consists of multi-disciplinary persons.

高端设计方法是一种设计哲学，也是一种将已有的设计技巧与人类科学、技术和商业领域的其他学科整合后的浓缩强化的设计方法。

High Design is a philosophy of design，as well as an enriched design approach that integrates established design skills with other disciplines in the areas of human sciences，technology and business.

1．项目管理

1．Project management

飞利浦设计有一套自己的项目管理的模式，部分重叠的双三角形管理模式。在客户满意三角形中，三个角分别是时间、金钱和质量；而“使能实现”三角形的三个角分别是信息、组织和沟通。

Philips Design has its own project management model：partly overlapping dual triangles. In triangle of client satisfaction，time，money and quality are on three corners，as well as in triangle of enabling information，organization and communication are on three corners.

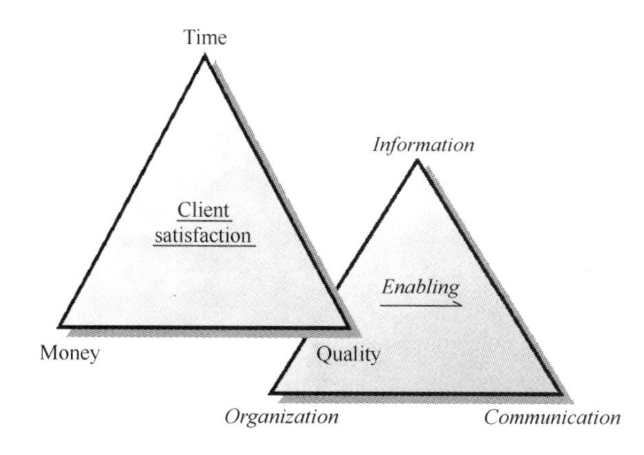

飞利浦设计公司的双三角形管理模式
Dual triangle management model of the Philips Design.

2．质量导向与质量评估

2．Quality lead & quality review

质量导向是设定要求并指引设计过程；质量评估是指将结果与要求进行比较。这意味着在设计的各个阶段要反复地进行模拟和评价，以不断改进设计过程。

Quality lead is setting requirements and directing process；quality review means to compare results with requirements．It means that during design phases simulations and evaluations should be implemented again and again in order to improve the design process continuously.

3．设计过程

3．The process

在高端设计中，整个设计过程被划分为五个阶段：初始阶段、分析阶段、概念设计阶段、定案阶段和评估阶段。初始阶段得出该项目的任务——商定项目的内容。分析阶段的结果是得出商定的设计要求。概念设计阶段的工作是为了得出协商同意的一个(组)设计概念。定案设计阶段的工作得到确定的设计规范和/或设计版本。评估阶段则是众多反省和学习时机，以确定新的改进机会。

In High Design，whole design process consists of five phases：initiation phase，analysis phase，concept design phase，finalization phase and evaluation phase．Initiation phase leads to the project assignment—agreed project context. Analysis phase results in agreed design requirements. Concept design phase works towards a(set of)agreed design concept(s)．Finalization phase works towards the defined design specification and/or design release. Evaluation phase is reflection & learning moments，identifying new & improvement opportunities.

对于高端设计，其核心设计阶段，即分析阶段、概念设计阶段和定案阶段，都基于反复使用高端设计的基本循环：分析、创新和评价。在这些阶段中如果需要，都可返回到前一阶段去进行修改可能的矫正行动(亦即由评估引起的改进活动)。

For High Design，the core design phases，i.g．analysis phase，concept design phase and finalization phase are based on the repetitive use of the basic High Design cycle：analysis，creation and evaluation．During these phases，there are possible corrective actions in a loop backwards to a previous phase if so required(e.g. improvement actions resulting from the evaluation).

每个项目都要确定其各个设计阶段及其规划和活动，要交付的东西、责任，以及使用的工具。

For each project will be defined design phases，planning and activities，deliverables，responsibilities，tools.

4. 工具包

4. The toolbox

对于每个设计阶段，高端设计法都有许多设计工具。例如，在分析阶段就有路线图、策略意图路线图工具、趋势和机会矩阵、可用性性能指标、利益路线图、概念创新工作坊、情绪看板法等工具可以使用。

The High Design has a lot of design tools for each phase. For example, in analysis phase, the tools, such as roadmappings, strategic intent mapping tool, trends & opportunities matrix, usability performance indicators, benefit mapping, concept creation workshop, moodboard and so on, could be used.

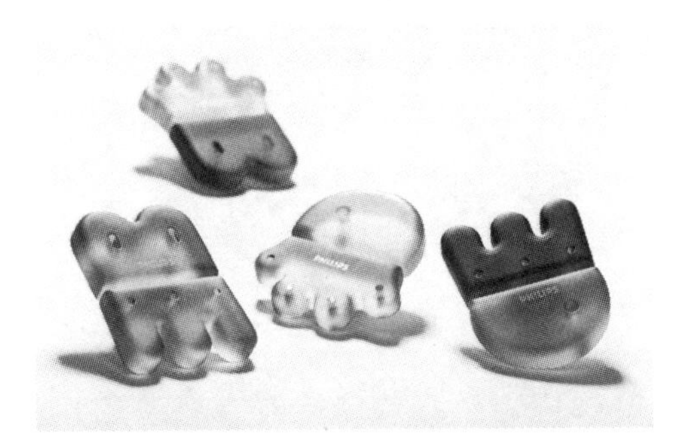

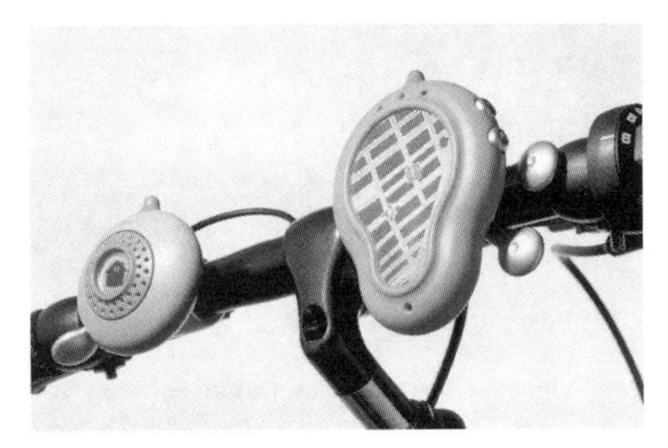

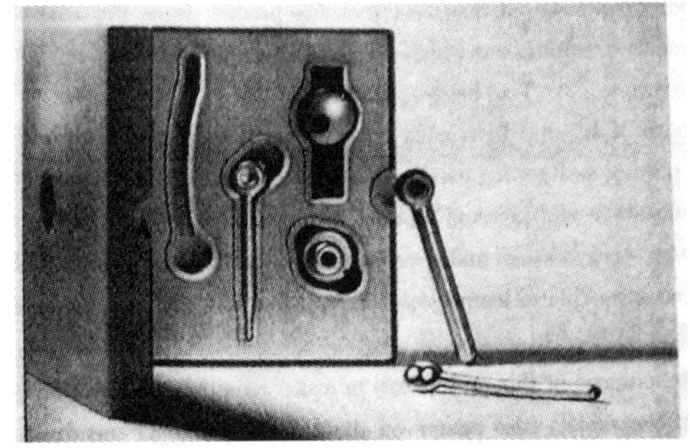

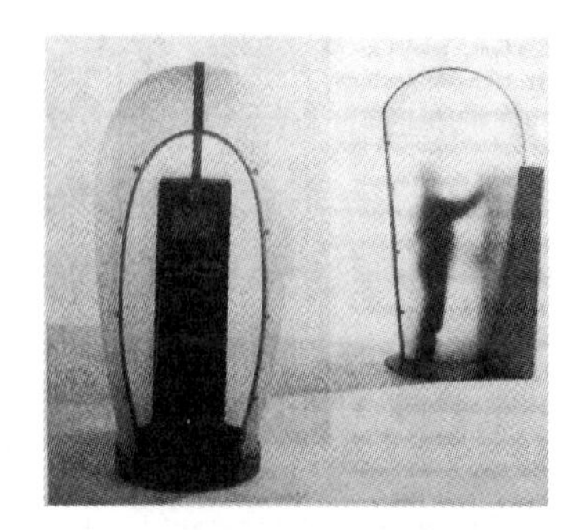

飞利浦设计公司在20世纪90年代发布的“未来设计”(按次序)：热徽章；情感容器；车载导航仪；自行车用游戏器；多媒体租赁；家用医疗盒；便携影像邮寄卡；多媒体亭。他们已经进入我们生活，甚至已经过时

“The designs for future” released by Philips in 1990s(according to orders): Hot Badges; Emotion Containers; In-Car Navigation; Biko Games; Multi Media Rentables; Home Medical Box; Moving Image Postcard; Multi-media Kiosh. They alReady came into our life, indeed some out-of-date.

拓展阅读

Extend and reading

21 Ways to kill an idea

(1) Ignore it.

(2) See it come and dodge.

(3) Scorn it.

(4) Laugh it off.

(5) Praise it to death.

(6) Mention that it has never been tried.

(7) Prove that it isn’t new.

(8) Observe that it doesn’t fit company policy.

(9) Mention what it will cost.

(10) We have tried before.

(11) Encourage author to look for better idea.

(12) Find a competitive idea.

(13) Produce 20 good reasons why it won’t work.

(14) Modify it out of existence.

(15) Encourage doubt about ownership.

(16) Damn it by association of ideas.

(17) Try to chip bits off it.

(18) Make a personal attack on the originator.

(19) Score a technical knockout.

(20) Postpone it.

(21) Let a committee sit on the idea.

Sony's and Toshiba's walkman

During the 1970s, when the walkman was first prototyped, tape recorders were products which played and recorded sound. Presented with a machine which not only lacked a recording facility but which also could be used by only one person, through headphones. Sony's marketing department concluded that this was "a dumb product". Fortunately, the chairman Mr Morito had tried the walkman himself and loved it, as he himself loved music as well as playing golf, it was his dream that when he played golf he could listen to the music. The marketing department's skepticism was over-ruled and a great success story resulted. The success market needs research started in asking question of: "do you like to listen to music while you are walking or moving?"

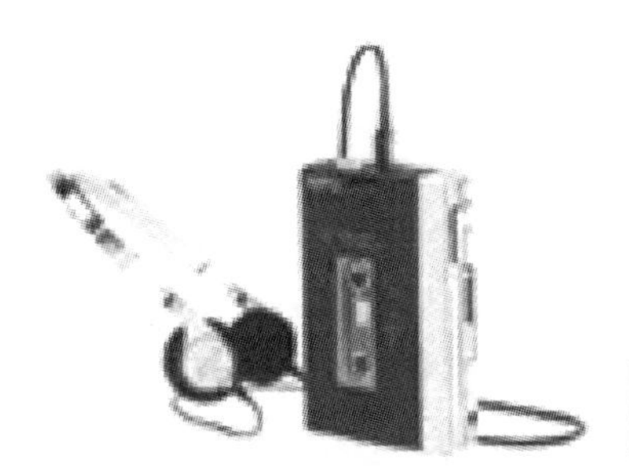

索尼设计和制造的该款随身听是世界上第一台
The walkman designed and manufactured by Sony is the first one in the world.

When Toshiba started to develop the walkman, they analyzed Sony's walkman in the market, which appealed for "high quality" and "high technology". So Toshiba decided to produce walkman with "high fashion", targeted on young people, especially the young girls, so that the Toshiba's walkman are also very successful.

第 24 课 | 以用户为中心的交互设计
Lesson 24 | User-centered interaction design

词汇和短语
Words and phrases

interaction　交互作用，互动作用
accuracy　精确度，正确度
efficiency　效率，功效
usefulness　有用，有效性
satisfaction　满意，满足，令人满意的事物
ownership　所有权，物主身份
relationship(with)　关系，关联
prototype　原型，样机
validate　使有效，使生效，确认，证实
rebut　反驳
persona　人，角色
scenario　情节
wireframing　线框图(法)
bug catching　缺陷捕捉

词汇联想与记忆
Association and memory of words

figure out　算出，想出，弄清楚……
(be) aware of　知道，察觉到，意识到

think of...as... 把……看做是……，
认为……是……
(be) thought of as 看做……

关键词汇和概念
Key words and concepts

交互设计旨在提高工作的正确性和效率而不减弱其实用性。该目标是要减少无效工作、提高使用者的生产率和满意度。

Interaction design aims to increase accuracy and efficiency of a task without diminishing usefulness. The objective is to reduce frustration and increase users' productivity and satisfaction.

交互设计，首先通过研究和理解特定使用人群的需求然后进行设计以满足并超越需求，努力改善产品的可用性和体验。

Interaction design attempts to improve the usability and experience of the product, by first researching and understanding certain users' needs and then designing to meet and exceed them.

课文
Text

以用户为中心的交互设计
User-centered interaction design

因为新技术对于意向目标受众经常是十分复杂的，交互设计旨在使学习的弯路减到最小，提高工作的正确性和效率而不减弱其实用性。该目标是要减少无效工作、提高使用者的生产率和满意度。

As new technologies are often overly complex for their intended target audience, interaction design aims to minimize the learning curve and to increase accuracy and efficiency of a task without diminishing usefulness. The objective is to reduce frustration and increase users' productivity and satisfaction.

交互设计，首先通过研究和理解特定使用人群的需求然后进行设计以满足并超越需求(弄清楚谁需要使用它，以及这些人喜欢怎样地使用它)，努力改善产品的可用性和体验。

Interaction design attempts to improve the usability and experience of the product, by first researching and understanding certain users' needs and then designing to meet and exceed them. (Figuring out who needs to use it, and how those people would like to use it.)

只有定期地把将要使用一个产品或系统的使用者拉进来，设计师才能够通过合适的修整使可用性最大化。把实在的使用者拉进来，设计师就能增强能力更好地理解使用者的目标及体验。还有一些积极方面的影响因素，包括提高对系统能力的了解及使用者主权意识。重要的是，使用者从较早阶段就对系统能力有所了解，从而对于功能性的预期有现实而适当的理解。另外，作为在产品开发中的一个积极参与者，这些使用者更会有拥有者的感觉，从而全面增加满意度。

Only by involving users who will use a product or system on a regular basis will designers be able to properly tailor and maximize usability. Involving real users, designers gain the ability to better understand users' goals and experiences. There are also positive side effects which include enhanced system capability awareness and users' ownership. It is important that the users are aware of system capabilities from an early stage so that expectations regarding functionality are both realistic and properly understood. Also, users who have been active participants in a product's development are more likely to feel a sense of ownership, thus increasing overall satisfaction.

与使用者界面设计的关系

Relationship with user interface design

交互设计常常是与各种媒体的系统界面设计有关，但是交互设计集中于界面在整个时间段上所限定和呈现出来的行为的方方面面，聚焦于开发该系统以响应使用者的经验而不是相反。系统界面可以被看做是人造物(无论是视觉上还是其他感觉上)，它象征一种按意愿计划的互动。(电话使用者界面中的)交互声音反响就是作为媒体而没有使用者图形界面的一个交互设计例子。

Interaction design is often associated with the design of system interfaces in a variety of media but concentrates on the aspects of the interface that defines and presents its behavior over time, with a focus on developing the system to respond to the user's experience and not the other way around. The system interface can be thought of as the artifact(whether visual or other sensory)that represents an offering's designed interactions. Interactive voice response(of telephone user interface)is an example of interaction design without graphical user interface as a media.

然而，交互性不限于技术系统。人类作为一个物种开始就已经彼此互动了。因此，交互设计可以用于开发所有的解决办法(或提议)，诸如服务和事件。一般来说，设计这些提议的人本质上已经完成了交互设计，虽然没有这样命名它。

Interactivity, however, is not limited to technological systems. People have been interacting with each other as long as humans have been a species. Therefore, interaction design can be applied to the development of all solutions(or offerings), such as services and events. Those who design these offerings have, typically, performed interaction design inherently without naming it as such.

方法论

Methodologies

对于一个已知的界面设计问题，交互设计师常常遵循相似的程序来创造一个解决方案。设计师建立几个快速原型并让使用者来测试它们，以肯定或反驳该理念。

Interaction designers often follow similar processes to create a solution to a known interface design problem. Designers build rapid prototypes and test them with the users to validate or rebut the idea.

在交互设计中有六个主要步骤。在用户反馈的基础上，这些步骤中的任何几步又都可以出现若干次的循环反复。

There are six major steps in interaction design. Based on user feedback, several iteration cycles of any set of steps may occur.

1. 设计研究

1. Design research

使用设计研究技术(观察法、面谈法、问卷调查法及相关活动)，设计师对用户及其环境进行调查研究，更多认识他们以便更好为他们设计。

Using design research techniques(observations, interviews, questionnaires, and related activities)designers investigate users and their environment in order to learn more about them and thus be better able to design for them.

2. 研究分析和概念产生

2. Research analysis and concept generation

采用使用者研究、技术可行性和商业机会分析的结合，设计师为新的软件、产品、服务或系统产生一些概念。这个过程可以包括多个轮回的头脑风暴法、讨论及推敲提炼。

Drawing on a combination of user research, technological possibilities, and business opportunities, designers create concepts for new softwares, products, services, or systems. This process may involve multiple rounds of brainstorming, discussion, and refinement.

为了帮助设计师了解使用者的要求，他们可以使用诸如角色法或使用者剖面法等工具，这些都是他们的目标使用人群的反映。从这些角色以及研究中观察到的行为模式出发，设计师创造情节(或使用者故事)或故事板，来设想使用者未来使用该产品或服务的运作流程。

To help designers realize users' requirements, they may use tools such as personas or user profiles that are reflective of their targeted user group. From these personae, and the patterns of behavior observed in the research, designers create scenarios(or user stories)or storyboards, which imagine a future work flow the users will go through using the product or service.

在使用各种工具和模型进行充分分析研究之后，设计师创造一个横跨各个层面的使用者要求的高水准的概要。它包括了有关某项目目前与未来目标的可视化的陈述。

After thorough analysis using various tools and models，designers create a high level summary spanning across all levels of user requirements. This includes a vision statement regarding the current and future goals of a project.

3．可供选择的设计与评估

3．Alternative design and evaluation

一旦问题空间清楚显现，设计师将用草图模型开发可供选择的解决方案，以帮助传达概念和理念。提出的解决方案要予以评估，甚至可能被归并掉。最后的结果应该是个解决了尽可能多使用者要求的一个设计。

Once clear view of the problem space exists，designers will develop alternative solutions with crude prototypes to help convey concepts and ideas. Proposed solutions are evaluated and perhaps even merged. The end result should be a design that solves as many of the user requirements as possible.

在这个过程中可以采用的工具是线框图法和流程图表法。一个产品或服务的特点和功能性常常用称之为线框图(另一种叫法是“图表法”)的文件概要地表达出来。线框图就是一页一页或者一帧一帧地把系统详尽地表达出来，它还包括该系统将任何运作的注释(“评注”)。流程图表法则将该系统的逻辑和步骤或者一个单独的特点概要地表述出来。

Some tools that may be used for this process are wireframing and flow diagrams．The features and functionality of a product or service are often outlined in a document known as a wireframe (“schematics” is an alternate term). Wireframes are a page-by-page or screen-by-screen detail of the system，which include notes(“annotations”)as to how the system will operate．Flow diagrams outline the logic and steps of the system or an individual feature.

4．原型及可用性试验

4．Prototyping and usability testing

交互设计师们使用各种原型技术来试验设计概念的各个方面。它们大体可以分为三类：试验人造物角色任务的，试验其外表和感觉的，以及试验其执行情况的。有时它们也被叫做体验原型，以强调其交互本性。原型可以是物理的或数字的，高保真度或低保真度的。

Interaction designers use a variety of prototyping techniques to test aspects of design ideas．These can be roughly divided into three classes: those that test the role of an artifact，those that test its look and feel and those that test its implementation. Sometimes，these are called experience prototypes to emphasize their interactive nature. Prototypes can be physical or digital，high- or low-fidelity.

5．执行

5．Implementation

交互设计师们在产品或服务的开发过程中需要参与其中以确保设计概念能正确执行。常常在建造过程中需要作些改变，交互设计师们应该参与在过程中对设计所作的修改。

Interaction designers need to be involved during the development of the product or service to ensure that what was designed is implemented correctly．Often，changes need to be made during the building process，and interaction designers should be involved with any of the on-the-fly modifications to the design.

6．系统测试

6．System testing

一旦系统建立，常常要执行对可用性和误差(“缺陷捕捉”)的另一轮测试。理想地讲，设计师们在此同样也要参与对系统作些必要的修改。

Once the system is built，often another round of testing，for both usability and errors (“bug catching”)is performed．Ideally，the designers will be involved here as well，to make any modifications to the system that are required.

拓展阅读

Extend and reading

Some knowledges about interaction design

Social interaction design

Social interaction design(SxD)is emerging because many of our computing devices have become networked and have begun to integrate communication capabilities．Phones，digital assistants and the myriad connected devices from computers to games facilitate talk and social interaction．Social interaction design accounts for interactions among users as well as between users and their devices. The dynamics of interpersonal communication，speech and writing，the pragmatics of talk and interaction-these now become critical factors in the use of social technologies．And they are factors described less by an approach steeped in the rational choice approach taken by cognitive science than that by sociology，psychology，and anthropology.

Affective interaction design

Throughout the process of interaction design，designers must be aware of key aspects in their designs that influence emotional responses in target users. The need for products to convey positive emotions and avoid negative ones is critical to product success. These aspects include positive, negative, motivational, learning, creative, social and persuasive influences. One method that can help convey such aspects is the use of expressive interfaces. In software, for example, the use of dynamic icons, animations and sound can help communicate a state of operation, creating a sense of interactivity and feedback. Interface aspects such as fonts, color pallets, and graphical layouts can also influence an interface's perceived effectiveness. Studies have shown that affective aspects can affect a user's perception of usability.

Emotional and pleasure theories exist to explain people's responses to the use of interactive products. These include Don Norman's emotional design model, Patrick Jordan's pleasure model, and McCarthy and Wright's technology as experience framework.

Interaction design domains

Interaction designers work in many areas, including software interfaces, (business)information systems, internet, physical products, environments, services, and systems which may combine many of these. Each area requires its own skills and approaches, but there are aspects of interaction design common to all.

Interaction designers often work in interdisciplinary teams as their work requires expertise in many different domains, including graphic design, programming, psychology, user testing, product design, etc.(see below for more related disciplines). Thus, they need to understand enough of these fields to work effectively with specialists.

History

The term *interaction design* was first proposed by Bill Moggridge and Bill Verplank in the late 1980s. To Verplank, it was an adaptation of the computer science term *user interface design* to the industrial design profession. To Moggridge, it was an improvement over *soft-face*, which he had coined in 1984 to refer to the application of industrial design to products containing software.

In 1989, Gillian Crampton-Smith established an interaction design MA at the Royal College of Art in London(originally entitled "computer-related design" and now known as "design interactions"). In 2001, she helped found the Interaction Design Institute Ivrea, a small institute in Northern Italy dedicated solely to interaction design; the institute moved to Milan in October 2005 and merged courses with Domus Academy. Today, interaction design is taught in many schools worldwide.

附录Ⅰ：全书各课拓展阅读的参考中译文

第1课

诺基亚2650型手机

诺基亚2650手机关键性的突破在于它特殊的折叠机构。这种手机并没有像它的竞争产品那样使用铰链，结果不管是在开或合的位置看起来都特别的一体化。外刚内柔的2650手机，合起来时坚韧耐用，打开后对于使用者的脸就如贴在枕头上般柔软。

2650手机的设计体现了一种高品质的均衡、匀称和简洁，使它看起来较少高科技和复杂设备的感觉，更多地显示出这是一款简单实用的产品。除了这种特殊的折叠机构外，2650手机比其他折叠式手机更简单，制造成本更低。

耐克Considered鞋

耐克Considered鞋性能优越，它结合了精细的造型和独特的环境保护功能，而未牺牲耐克对设计创新的承诺。这款手工短靴炫耀它独特的外观，这得益于在皮革鞋面上编织纤维饰带以使鞋与脚的轮廓服帖。这个产品项目来自于消费者要求更经久耐用产品的反馈。设计师们达到了令人印象深刻的环境统计数据：产品生产过程中废物减少了61%，能量消耗减少了35%，溶剂使用量减少了89%。另外，这款鞋设计成可更有效地参与到耐克的旧鞋回收利用计划中去。这个产品投放市场的效果超出了预期，并促使耐克公司考虑在其他产品生产线中采取类似的办法。

“这是令人惊异的高技术和手工艺的结合。”

“这双鞋真是个大赢家，它达到了所有设计的和谐：好看，富有人情味，有手工艺的感觉，符合生态，功能完备，还很时髦！耐克将燕麦片和花椰菜转变成了时髦鞋”。

iXi自行车

目标是用跟其他时尚耐用品同样的方式设计一款视觉上充满魅力，并与使用者友好的自行车。它应该吸引通常可能不会冒昧进入传统自行车商店的那些世故消费者。同时，这种自行车不仅在骑乘时，而且在停放时都能方便地使用，轻而易举地融入主人的日常生活。整个传动带系统根本无需维护，无需张紧和润滑。媒体在该车面市后的关注大为提升了公司的形象，结果销售更顺畅，并使整条生产线在业内获得了重新关注。

RKS流行系列吉他

该流行系列吉他是个高质量、充满革新意识、功能全面的乐器，拥有多种抢眼的色彩。其设计策略注重两个方面：近年来吉他行业中缺乏创新，及吉他生产过程中不可替代自然资源的消耗。这个流行系列吉他在人机工学和声质方面达到了完美的平衡。以许多大胆而流行的色彩对琴杆和琴身进行各类搭配，生产出了多种多样风格的产品，吸引了范围广泛的各类消费者。从持续性的角度看，这种吉他所用的高音质木材的总量明显减少。普通吉他要用8～10磅来自于中南美洲热带雨林的高音质木材。而RKS流行系列吉他仅用不到2磅本地产木材。

第2课

影拓3：计算机设备金奖

影拓3手写板系统是一种无线即插即用型、不需电源的、供专业人士使用的综合电脑图形工具。Wacom公司希望有一款创新的、易接近并符合人机工学的设计，它应在视觉上清晰传达其使用方便性，并因此赢回公司输给竞争对手的市场份额。广泛的观察与人机工学研究为产品创新确定了许多机会，例如工作适应性、直觉的功能性、改进了的美学以及逼真的压延成型与着色模拟技术。

新的设计给使用者提供了他们为捕捉灵感所需的控制力、精确度和灵活性，其人机工学设计可使他们超长时间地绘图和着色而无不舒感。从2004年上市以来，影拓3的销售超出了人们的预期，产品的销量是同类竞争者的两倍，帮助公司的利润率提高了近11%。

索尼Qualia 016数码相机

索尼Qualia 016是一款极其紧凑的掌中数码相机。其尺寸使该相机能与你始终相伴，随时进行拍摄。Qualia 016不仅仅只是一台照相机，并且有一整套附件来优化你的摄影体验，诸如取镜器、广角镜、望远镜、闪光灯、视频输出单元和定时遥控单元，所有这一切都安放在一个定制的便携包内。然而，其设计创新远不止是其尺寸较小。设计师使得相机的界面变得更舒服、更直观，总是成为您手的延伸，并成为公司在创造吸引人心的产品时的焦点。

“这是每一个装备良好(及穿着考究)的收集迷都想要的圣诞礼物。这台完美精致的超小型数码照相机，只有你两个手指的大小，从精细加工的机壳、镜筒和许多附件一起装在一个小巧的包中，它们的美吸引着你。拍照时只需用两个手指之一沿着长条形接触件选择即可。”

弹性抓筒——逗狗玩具

这个注塑成型、用天然橡胶材料制成的逗狗玩具是为宠物和人共同设计的玩具。这种玩具本身壁厚不均，可产生无规则的弹跳和翻滚，而奖励物品可放进可塞东西的空心中。该玩具内核较厚，狗狗可以在更简单自然的位置抓咬住它。设计过程还关注到销售时对顾客的吸引力：包装简化到缠绕，以突出设计本身。这是WETNo2国际公司首次冲击狗玩具市场。一年的市场反响导致在2005年末增加新生产线投放市场的计划。

“谁能想到逗狗的玩具会是个彩色雕塑件？只要满足弹性十足，滚动自如，经久耐用，便于狗抓咬这些基本条件，设计师想‘为什么不让它形状和色彩都好看呢？’其结果是令人欢快的出乎意料，而不是到处都有(且有点丑陋)的生牛皮制骨头。”

第3课

奢移的小玩意

1．飞利浦的斯坦·泰斯拉手表

你是否担心过你应用的所有技术都可能损害你的身体？我们没担心过。但是如果你在意那看不见的“脑电波”，你必将钟情于这款手表。手表里有块泰斯拉芯片(是以那位曾和托马斯·爱迪生共事的发明家的名字命名的)，它发出一种信号以屏蔽电子污染，让你无忧无虑地享受技术成就。帕夫爸爸有块这种表，但我们还不能确认这到底是不是个好东西。

2．保罗·斯密斯护鞋用品套装

除了你的脸以外，你的鞋就是人们最关注你的第一件事。即使你是个帅小伙，你邋遢的鞋也会让你威风大减。使用这套包括两个上光刷、一块擦鞋布、三管不同颜色的鞋油、一个鞋拔和一个刷子的套装护鞋用品，就可以避免这种社交场合的尴尬。所有这一切都装在一个设计得看似装画家的颜料包的袋儿中。

3．德·塞得DS-151型椅子

我们认为你最好是坐下。不，我们并没有什么令人沮丧的消息，我们只是想让你试一下这个令你惊奇的、极其舒服的椅子！它有一个可调节的靠枕，它的支架在你劳累了一天坐进去时是不会在地上滑动的。最好之处，椅子的造型从人机工学上照顾了人们躺在里面双手抱头、绝对无所事事的动作。它太适合我们了。

4．ZARA3赛型婴儿车系列

ZARA3童车一个产品就可以满足搬运三岁及以下婴儿的所有需要——从婴儿汽车座到传统的婴儿推车，以及婴儿学步时的车座。这个合为一体的设计节省了父母的开支，其复杂程度被降到最低，使日常使用以及不同阶段间的变换很方便。设计师去掉了所有为增加功能所需的多余零件。底盘紧凑，使得它能轻易举起并放进汽车里。对于顾客来讲，布料织物具有广泛的美学选择余地，也成为用户将其与其他产品进行区分的一个方法，并在产品使用寿命期内可随意更换织物件以及用布料织物来代替一些机械零件。例如，通过调整靠后背和脚蹬架，手推车可以有两个位置，而这只需要借助拉链通过织物、维柯牢尼龙搭扣和夹子等就可实现。

第4课

艺术，设计以及格式塔理论

撰稿：罗伊·R·贝伦斯

摘要：格式搭心理学是在1910年由三个德国心理学家创立的，他们是马克斯·维梯莫，库特·库夫卡和沃尔夫岗特·库勒。本文作者论述了格式塔理论对现代艺术和设计的影响，阐明了它与那些受日本理论激发的美学理论之间的相似之处，找出了格式塔心理学家与某些艺术家之间存在共同兴趣的证据，即便很有限。

(摘录)

格式塔心理学1910年起源于德国。在一次坐火车度假旅行时，一个生于捷克的30岁心理学家马克斯·维梯莫当他在一个铁路交叉道口看见闪烁的灯光，恰似剧院里大屋顶下的一圈圈灯光时被一个念头吸引住了。他在法兰克福车站下了车，买了个被叫做“西洋镜”的运动图片玩具。当放在西洋镜里的一系列图画纸片被抽动、从缝隙里看时，连续的固定画面看起来就成为一个活动的图画。在旅馆的房间里，他制作了自己的画片条，里面不是可识别的物体，而是些简单抽象、从垂直到水平间变化的线条。通过这些元素的变化，他能研究影响这些运动图片幻影的条件，这就是技术上称为“显动作”的效果。

早几年前，维梯莫在布拉格师从奥地利哲学家克里斯汀·冯·艾伦费尔斯学习，后者在1890年发表了一篇名为“格式塔品质”的文章，在此文中他指出，用不同的琴键弹奏的曲子依然可以听出曲调来，虽然没有一个音符是相同的；他还指出，从一个诸如“方形”或“有棱角的形状”的特征里抽象出来的形状，可以用范围很宽的特定元素来传达。艾伦费尔斯论述道，显然，如果一个曲调和组成此曲调的音符是如此的互不相干，则整个曲调就不是简单地将其各部分加起来，而是互相促进的“总效果”，或称为格式塔。艾伦费尔斯下结论说，显运动的效果主要不是由个别元素而产生，而是由它们间的相互动态关系发生。

……

格式塔心理学的三个奠基人由于第一次世界大战而分开，在1920年又汇聚一起，当时库勒成为柏林大学心理学学院的院长，而维梯莫则已经就职于该学院。他与库夫卡一直保持着联系，当时库夫卡还在继续法兰克福附近教书，而维梯莫和库勒则在一个废弃的皇宫里建立了研究生课程，并开始出版一本称为《*Psychologische Forschung*(心理学研究)》的研究性杂志。大部分时间，学生并不去听讲课，而是用学生作为课题的研究人员进行实际指导的研究，并为出版准备论文。这种方式成功的证据就是，我们现在耳熟能详的许多心理学名字就是该学院里的教师和学生，例如……(简略——本书编著者)。

库夫卡在1924年离开欧洲到达美国；维梯莫则在1933年到达美国。在1930年代早期，心理学研究所开始衰退。当国家社会主义在1933年开始掌权，在他们的直接威胁行为中就包括解雇大学里的犹太族教授，从诺贝尔奖获得者到研究助理。同情“犹太人维梯莫”的流言开始传播，库勒公开遣责反犹太主义，并在柏林的报纸上发表文章抗议解雇犹太人，该文是纳粹容忍的最后一篇这样的抗议文章。连库勒自己也感到吃惊的是，他没有被捕，但受到的威胁加剧，他在1935年也移民到了美国。

格式塔心理学家中间没有一个是艺术家，也极少有设计师，但很早就有这两个学科互相感兴趣的征兆。比如1927年，格式塔心理学家鲁道夫・安赫姆就访问了迪绍・包豪斯，然后在*Die Weltbühne*上发表文章称赞了包豪斯建筑设计的坦诚和透明。不久，格式塔学者库特・刘因委托彼得・贝伦斯(包豪斯奠基者沃尔特・格罗皮乌斯的老师)设计他在柏林的家，但是后来由于意见不和，邀请包豪斯的家具设计师马歇尔・布鲁埃尔完成了室内设计。1929年由于行程安排上的冲突，库勒谢绝了包豪斯请他去讲学的邀请，而让他的学生卡尔・当克尔代替他前往。在听众中有画家保罗・克利，他早在1925年就知道了维梯莫的研究。但其他的包豪斯艺术家也是兴趣浓溢的，包括韦斯利・康定斯基和约瑟夫・奥尔伯斯，两人还出席了1930～1931年间由来自莱比锡大学的访问心理学家康特・卡尔佛利特・冯・杜克海姆所作的格式塔系列讲座。

实在讲，艺术家信奉格式塔理论的一个原因是，它为有关成分和页面布置的古老原则在精神上提供了科学的根据。格式塔理论在法语中有个绰号叫印版的心理学。由于强调绝对抽象的形态、结构的经济及暗示，格式塔理论渐渐和现代主义一起趋向唯美主义，如同音乐和建筑一样信仰所有的艺术本质上都是抽象的设计，正如艾伦・卢布坦和J・阿博特・米勒在《设计写作研究》(1966年)所表述的那样，“设计从根本上说是抽象的形式活动”而“文本(或主题内容)是次要的，只在掌握了形式之后附加上去的东西”。

第5课

圈形椅

这两项是室内家具和室外家具神奇的混合物。它们都是在以使用石棉水泥著称的瑞士，以手工制作完成的：没有石棉的淡灰色光纤水泥。这是一把无论在游泳池边还是空大的阁楼上都令人惊奇的椅子，实际上它还是把摇椅。它重量很轻，移动方便，是理想的休闲坐具。圈形的桌子带有两个凹下去的饮料座，与环形椅底恰到好处地配合便于存放。它们极其舒服，是对速度型产品的重大添加。我们钟爱这两个东西的外观和感觉，认为把它们加进我们的资产清单里是非常理想的，在漫漫夏夜、在游泳池边，有了它们就可以惬意地享受醉人的鸡尾酒了。

这种传奇的沙滩椅是由威利・高尔在1954年设计的，是现代家具设计的杰作。在费拉德尔菲亚(美国费城)艺术博物馆、韦德拉设计博物馆和博伊德收藏品的收藏中，1998年威利・高尔本人的重新设计再次证明了设计师的信条——“用最有小的努力实现最优化”。

圈形椅是非常出色的，因为它是把独立而无需另外支撑的纤维水泥座椅。无论放在室内还是室外，这椅子看起来都像是雕塑和艺术品。作为无石棉的水泥纤维的结合，具有光滑、温暖的表面，几乎不会轻易损坏。它在瑞士由手工制造完成。

第6课

项目：48英寸柔墙

设计者：穆罗

那么完美的游戏之屋！纸的柔墙是很美，质量很轻，可以自由直立的墙，能够摆成几乎任何形状，或很容易地被卷成紧凑的一束储藏在任何地方。这种柔墙能够吸声，吸收和传导光线。这种纸质柔墙用400层蜂窝状半透明的耐火白纸制成，以天然羊毛毡收口。当墙面展开时，厚厚的毛毡边成为把手；当墙体压缩时，毛毡又成了保护的外壳。柔墙是标准尺寸，毛毡收口的地方有维尔克洛搭扣，可以让它们连接在一起成为一堵墙。纸质柔墙是精致的，蜂窝状结构设计保证了正常操作下它有超强弹性。

柔墙能够一个人展开，虽然初次打开时两个人做更容易些。柔墙在光滑、洁净、干燥的表面上能够滑动时移动它最好。不使用时，可以平放保存或对半折叠保存。

柔墙不是为支撑重物或任意变形设计的。这种纸质软墙可以轻轻掸尘，或用吸尘器的毛刷头轻轻除尘，但不允许弄湿。软墙的纸体经过阻燃处理，本身不会有明火；但这种纸质软墙不可放靠近明火或热源处。如果纸张发生撕裂，纸质墙体能够隐蔽或保护受损区域。由于其两面相同，软墙可以正反使用。

柔墙四英尺高，它可以用来分割空间，或者哪怕仅仅通过部分阻隔视线的方法，无论站着或坐着都能创造单独空间的感觉。软墙的厚度是12英寸(30厘米)，且能以每1.5英寸(3厘米)的间隔增厚到25英尺(7.5米)。纸为白色，中灰色的天然羊毛毡收口，重量只有10磅(4.5千克重)。

项目：Uten.Silo(小器具)

设计者: 多罗蒂・贝克和英果・茂热

Uten.Silo是20世纪60年代晚期最知名的塑料制品设计。今天，这种色彩丰富的挂在墙上的杂物收藏架(袋)已经成为真正的设计标志，其完全忠实于原作的新版产品现在已经由维特拉设计博物馆重新发行。

在1960年代末，塑料正成为占统治地位的设计材料。在意大利，诸如乔·科伦坡及维柯·梅杰斯特雷蒂这样的设计师都用它设计了鲜亮和令人愉悦的家具、照明设备和日常消费品，并由诸如阿特米德与卡特勒这样富有革新精神的生产商生产。以博芬格椅以及潘通椅为代表的、完全由塑料制造的椅子也诞生了。与此同时，在慕尼黑，早以其“灯泡”照明而引人注目的英戈·梅勒推出了这个引起轰动的塑料挂壁式杂物袋Uten.Silo。

Uten.Silo是由梅勒的妻子多乐蒂·贝克设计的。有着不同形状和大小的口袋，及其金属的挂钩和夹子，Uten.Silo可以用来组织收纳办公室、厨房、卫生间和儿童房的各种杂物。这个设计做到在工业的精确度和十分有趣的变化多样之间以及在合乎逻辑的组织和富有幽默的设计之间的张弛力，使得Uten.Silo成为一个功能强大的设计，也使塑料得到了恰如其分的使用。Uten.Silo的主体为ABS塑料，并遵循设计原型安装了金属挂钩。

第7课

贴点法

这是一种将许多利害关系人组织起来参与选择概念的好方法。它简单易行，无需提供选择的理由或动机等的大量反馈信息。

贴点法是一个好的筛选工具，适合于潜在方案或竞争方案很多，以及参与选择的利益相关者很多的情况。分给每一个参选者一定数量的贴纸“点”，他们可以把贴纸贴在一个点、两个点或贴到他们中意的全部点上。

方法

1. 确定概念

确保提供的每一个概念都得到同等程度的阐述，并保证给一定素质参选人员足够的信息，从而得到合理的选择。

2. 确定参选者

谁来做选择？只是内部利益相关人员中的代表性样本，或者把外部的利益相关人也可介入？

3. 分配贴纸“点”

给每一个参选者一定数量的贴纸“点”。根据所选概念的数目，分配的“点”数可以是3～5个。不同颜色的贴纸“点”可用于不同的用途。每一个参选人员可用不同颜色的“点”来表示他或她最满意的选择——这对平局决胜时会很有用。也可以把不同颜色的“点”分发给不同职能部门的代表或不同市场细分中的消费者，用来观察是否存在观点上的偏见或偏好。

4. 投票和筛选

第一次投票后，可以筛选出结果并进行第二轮投票，第二轮里只留下那些最强硬的竞争者。

5. 捕获喜好的憎恶

捕获隐藏在选择后面的理由很重要，这可以让参与者在贴纸上解说喜好和憎恶的理由，并可附上不同的解决方法。这将有助理解为什么某些方案比其他一些方案更受欢迎。这也能使被拒绝的方案中的优秀特征得以保留下来。

实例：概念和方案选择

概念和方案选择的目标是为了从广泛的设想概念和方案中确认哪些是解决最初设定的问题的最优方案。

每个想法是否已讨论并对照支持或拒绝的根据进行过权衡？	5 4 3 2 1	1 2 3 4 5	小组是否善待那些提出的想法受到批评或被拒绝的人？
小组在使用选择标准时是否系统化？	5 4 3 2 1	1 2 3 4 5	小组是否集中精力在选择最佳方案上，而非集中于拒绝不好的方案上？
小组对初步想法作了怎样的改进和整合？	5 4 3 2 1	1 2 3 4 5	不同观点是否商讨过并达到了共同的满意点？
为了更彻底地探索和评估，是否选好了一个(或一套)方案？	5 4 3 2 1	1 2 3 4 5	所选的解决方案是否是一致同意的？如果不是，小组内多大程度上同意？
概念产生讨论时的规则是否一开始就检查过和同意了？	5 4 3 2 1	1 2 3 4 5	是否不管其内容如何，所有的想法均得到认同和受欢迎？
是否每个人到最后都为概念产生竭尽全力了？	5 4 3 2 1	1 2 3 4 5	是否鼓励了少数新来的小组成员？
所有方案出台后，是否为了清晰、详尽和补充完善而在组内复审过？	5 4 3 2 1	1 2 3 4 5	批评意见是否被巧妙地劝阻，评估是否被有效地推迟？
是否所有方案已按特性或按属性分类？	5 4 3 2 1	1 2 3 4 5	个别人主导讨论或将其想法强加给小组的情况是否予以防止？
最具创新的、可行的或有趣的方案的汇总表是否已列出？	5 4 3 2 1	1 2 3 4 5	所有方案是否已陈述或贴出来供大家阅知？

第8课

阿莱西的设计师：亚历山德罗·曼蒂尼

1931年生于米兰，Domus(意大利著名的建筑和设计杂志——编者注)的前主管，Compasso d'Oro(今圆规奖，欧洲第一个设计奖——编者注)奖获得者，他是个设计师、建筑师，以及飞利浦等公司的图像顾问。

“亚历山德罗·曼蒂尼对我一直有点像个良师益友。当人们问我曼蒂尼为我们做什么、他的角色是什么时，我只是一笑。亚历山德罗·曼蒂尼的咨询作用包含了如此宽的方面，以致他的位置不可能实实在在地予以述说，也无法为外人所理解。”

“作为一个设计师，他持续地为我们提出目标，而这常常被发现是我们公司产品目录里最困难和最刺激的区域。作为一个建筑师，他为我们设计了阿莱西博物馆等。作为一个设计经理，他对于构思和总揽我们若干有声誉的设计项目而言是称职的。作为一个撰稿人，他为我们写了不少书。作为一个顾问，他向我推介了大量一直在为我们工作的设计师。作为一个亲密朋友，他明白我最隐私的问题和渴望……多少年来我们达到了一种工作上的密切配合，一种几乎是心灵感应式的关系……”

有关我自己的一些问题

我性格中的主要特点：积极向上。

我对男性品质的要求：复杂性。

我对女性品质的要求：复杂性。

朋友中我最欣赏什么：正派。

我的主要弱点：自我欣赏。

我最喜欢的职业：思考。

我的幸福梦想：不思考。

对我来说最大的不幸：世界末日。

我最喜欢成为的人：圣人。

我最喜欢居住的国家：我居住的地方。

我最喜欢的颜色：粉红。

我最喜欢的花：玫瑰。

我最喜欢的鸟：啄木鸟。

我最喜欢的作家：尼采。

我最喜欢的诗人：泰戈尔。

我幻想中的英雄：米老鼠。

我幻想中的女英雄：阿丽斯。

我最喜欢的作曲家：舒伯特。

我最喜欢的艺术家：塞维诺。

我真实生活中的英雄：格雷戈里·派克。

我心中的历史女英雄：英格兰维多利亚女皇。

我最喜欢的名字：耶稣十二使徒的名字。

我最憎恨什么：暴力。

我最不喜欢的历史性人物：独裁者。

我最崇尚的宗教改革：佛教。

我最喜欢得到的自然馈赠：随处可见。

我希望的死亡方式：在我的床上。

我目前的灵魂状态：稍有不安。

我能忍受的错误：所有的。

我的座右铭：不确定。

第9课

恩佐·马里——产品和家具设计师(1932-)

他是20世纪晚期最富有思想和智慧、最具有挑战性的意大利设计师之一。恩佐·马里事实上已经证明是对青年一代的设计师最有影响力的人物，同时他也是作家、教师、艺术家以及产品、家具和智力游戏的设计师。

他于1932年生于意大利的诺瓦拉，1952～1956年他在米兰博雷拉科学院学习文学和古典语言。作为一个学生，马里通过作为一个视觉艺术家和自由作家研究者的工作来保障自己的学习。当时有一段时间里，意大利的设计繁荣昌盛并带动了实业家同设计师的紧密合作，以此振兴他们的事业，他这时也开始对设计感兴趣，并刻苦地自学。

马里的设计方法着重于理论。他更在意设计在当代文化中的作用以及设计与使用者的关系，而不仅仅是成为一个设计从业者。1956年毕业以后，他在米兰自己成立了一个工作室，继续进行着视觉心理学、感知系统和设计方法论的研究。这些研究体现在马里所创造的线性元素和位面的三维模型中。为生活所迫，马里和意大利塑料产品生产商丹尼斯进行接触，并同意开发一系列大批量生产的产品。

16个动物是他和丹尼斯开始合作的第一个项目，于1957年上市。这是一个木质的拼图游戏板，马里应用他的解决问题理论，创造了一组由木头切割出来的简单动物形态，有河马、蛇、长颈鹿和骆驼等，拼在一起就是一个长方形。这个拼图游戏板标志着马里和这个公司长期合作的开始，一直继续到转入20世纪60年代开始生产容器和花瓶。马里在决定开发这些大批量生产产品的同时，并没有对于他的信念进行妥协，他始终认为：每个设计项目的成果都必须看起来和触摸起来都是很美妙的，同时又能有效地实现它的功能。他将他的这种哲学描述为一种“理性设计”，他认为他的设计“都是以一种与目标和功能完全协调的方式精心构筑出来的”。

马里在其他视觉艺术领域继续着他的实验工作，特别在1963年在米兰成立了称为Nuova Tendenza的艺术家小组。当然，作为设计师他也是产量颇丰。1962年，他开始了一个长达六年的开发项目，用丹尼斯公司的标志性材料——塑料——来开发衣帽架、伞架和垃圾筒。到1960年代末，马里已能巧妙、老练地处置塑料制品的设计，经他的手塑料能达到雕刻般的动感。他最成功的塑料产品设计之一就是编号为3087的花瓶，丹尼斯公司在1969年投入生产。这个花瓶中心有个圆锥体，使它无论是正过来或者倒过去站立都能使用。对于许多设计批评家而言，用光滑的ABS塑料制造的3087 花瓶具有如此诱人的形状，在对公众宣传塑料制品不一定是便宜的和俗气的这一点上起到了决定性的作用。

作为产品设计师他继续工作的同时，还转向了家具设计。1971年马里为德里艾德公司设计的Sof Sof椅子揭开了面纱，该椅子在简单焊接的杆形框架上只布置了一个可以拿掉的垫子。同样精致的还有为卡斯特力公司设计的1975-76盒子，这是一个买回去自行组装、由聚丙烯注塑模具成型的座位及环状金属管架子组成的椅子，它们可以分开后装在一个盒子里，就像拼板游戏块的互锁结构那样成为一个长方形一样。

在21年纪之初，马里已经到了70来岁了，他接受了日本无印良品(Muji)——一个注重理性唯美主义的日本家居用品商店的委任，并接受了维也纳的 Gebrüder Thonet的邀约创作了一个19世纪晚期著名的弯木椅的现代版本。其间，恩佐·马里从博雷拉科学院毕业不到一年就开始为其工作的丹尼斯公司生产了第他们一次合作成果——木制16个动物拼图的有限版本。

第10课

项目案例c：

李维斯的电子衣服LCD+的投放市场

客户的挑战：LCD+是市场上第一款消费者可穿用的电子衣服。这款衣服由飞利浦设计公司和著名的服装公司李维斯共同设计、飞利浦设计公司与飞利浦研究院共同开发的，在2000年投放市场。

投放市场的活动由飞利浦设计公司负责，对其的挑战在于要传达这样一种理念：不仅服装是创新的，而且也是整个新的生活方式的开端。

飞利浦设计公司的解决方案：投放市场活动的程序要允许观众同时体验新的生活方式和新的产品。

通过图像和声音采用了可感知的多媒体手段的展示使人们联想到舒适和自由。在媒体编辑们有机会试穿衣服的同时，有关技术和时尚相结合的历史讲座又陈述了这种理念，他们也就“不能自拔”了。

结果：对该产品的狂热是无法抵御的，投放市场活动在引起媒体关注方面具有雪球效应：仅在几个月内就有几百篇文章报道，而同时LCD+已经在着手其下一代。

关于飞利浦设计公司

我们是一个全球性的团体，任务是充实设计过程并向我们的客户提交具有竞争性的价值。我们创造的设计解决办法关注于个人的成长，使人们能够彼此和谐，并同他们的环境和谐地生活。

我们有数百人，分布在世界上的12个地点。这给我们以迷人的、显现地方趋势和发展的洞察力，其中有些可能产生远远超出他们地域的重要结果。这也使我们成为世界上最大的设计工作室之一。

我们信奉技术，但只作为达到目标的一种手段。这是一种启动力，是达到更好生活质量的一条途径。我们充满幻想的方法——用人类的科学、对使用者的研究以及总是明确地关注人来丰富设计——允许我们塑造技术以回应人们现有的和潜在的需求。

我们设计方案的目的是要使技术真正用以改善人们的生活质量，并使他们快乐。我们对所做的每一件事情都要坚持不懈地强调要基于明确的使用者选择，对于事物的关联性和来龙去脉以及真实诉求的需要。创造一个“快乐目标的前景”是可能的吗？当然能。

那么技术是不是确实在变得太烦人了呢？只有在我们让它这样时才是这样。从20世纪90年代初从事环境智能项目开始，我们总是试图冲击平衡，因而人们因技术创新而获得权力——但没有压倒优势的权力。

当然，我们相信我们的设计方案应该是负责的并且是可持续发展的。不论从商业、环境、个人、社会还是道德的观点看，确实没有任何其他切实可行的选择。

第11课

青蛙公司与灵活电子公司

灵活电子公司是谁？

灵活电子公司是最重要的电子制造服务(EMS)提供者，它关注对技术公司交付运作服务。灵活电子公司在2004财政年度的营业收入为145亿美元，是个重要的全球运营公司，它在32个国家与五大洲具有工程、制造和物流等方面的业务。这种全球化的存在允许它通过处于关键市场和地点的设备网络达到卓越的制造水平，从而用其资源、技术及其优化的运行能力提供给客户。灵活电子公司具有提供终端到终端的运行服务的能力，包括：创新的产品设计，试验解决办法，制造工艺，信息技术专门经验，网络服务及物流，已经确立了公司作为领先的EMS提供者的地位。

“……我们又一次跳到了曲线前面。” ——哈特姆特·埃斯林格

为什么是青蛙公司与灵活电子公司？

青蛙公司加上灵活电子公司将创造一个独特的、一应俱全的产品开发外协者；它将提供一种模式及使用者体验，而消费者就可能一起“坠入情网”，我们还将把这些产品更有效、省时省钱地推向市场。

“一起来吧”

青蛙设计公司的首席执行官之——哈特姆特·埃斯林格与灵活电子公司的首席执行官米歇尔·马克思共有一种观点：今天与明天的产品必须有益于提供卓越的消费者体验的所有玩家，从零售商到品牌拥有者、分销商到制造商。现在青蛙公司与灵活电子公司一起，能够整合从头到尾的整个PLM循环而不会危及他们的核心能力：世界领先的设计和创新，加上世界领先的技术、生产和物流。

青蛙公司通晓创新

30多年来，青蛙公司已经以其生产的市场领先产品、服务和品牌增益活动反复地显示了自己的卓越。我们的创造性办法已经使许多公司在竞争中迅速改变了市场地位。

青蛙公司懂得商业

我们独特的多学科交叉、并行创新过程是特别为我们客户的商业需要量身定制的。我们懂得商业面临的挑战，而我们的创新过程允许我们达到空前的投入产出比及上市时间。我们通过机械创新、数字创新和情感创新来造成商业冲击。

机械创新

我们通过创新帮助我们的客户戏剧性地增加投入产出比与品牌权益，适时地投放产品。我们在产品设计方面为苹果(麦金托什)、索尼(单枪三束彩色显象管电视机)和福特(“思索”车辆)所做的工作是传奇性的，它们带来的商业冲击是不容置疑的。

数字创新

今天的消费者强烈地需求数字存在。我们为戴尔、微软视窗XP及i2的工作证明了我们在使用者界面方面的经验。我们能够使其为你工作，通过有效执行电子商务策略而为你增加利润。

情感创新

我们的品牌团队能够通过我们已经过验证的品牌战略过程，帮助你的企业增加消费者忠诚度。我们曾为讴歌、甲骨文和SAP公司做过，我们也能为你做到。

第12课

乔纳森·艾夫

如果在苹果公司神圣的殿堂里加上一点噪声给艾夫·乔纳森的名字，他可能会被认为有点自大。其实这个留平头、说话温和的英国人完全不是这样。只有在他为帮助某些产品出名而给它们命名iMac、iBook、iPod时，你才会听到他说“我(I)”这个词。

艾夫一直努力把荣誉给予他组建起来的设计团队，他的指纹已经遍布在苹果五年多来硬件设计的重大转变上。在艾夫1992年被加州库柏蒂诺市的一家计算机制造商雇佣的时候，制造的产品还是浅褐色机箱的台式机以及近乎黑色的塑料笔记本电脑PowerBook。而当1997年斯蒂夫·乔布任命他为工业设计副总裁时，一切都改变了。

艾夫开始利用以前在工业中从未见过的材料、形态和色彩。最初的iMac用曲线、蜜糖色并带有携带手把，设计突破了浅褐色的箱体模式。没有别人尝试过建造像最新的iMac那样的一个计算机，在一根可移动的柱子上安置一台平面屏幕。没有别人用钛或者航空铝制造只有1英寸厚的膝上型电脑。还有，现在大家知道的用在最新的笔记本电脑PowerBook上的键盘，它会在知道天黑后自动点亮。

艾夫并不激进，他只是为细节而烦恼。“你涉及的是产品的需要和问题”，他说：“你常常得不到预料的结果”。与有些设计师不同，艾夫总是使用他自己设计而制成的产品。这样做给了他以后进一步改进的灵感。就像为iPod上的可移动轨迹轮做的敏感微调，使之轻触即可动作，这样你慢条斯理时它不会也慢条斯理。妙计就是如此成功，虽然它们是不可想象的复杂，但我们几乎不会注意。你可以对低调的艾夫有同样的评价。

传记

1967年　生于伦敦，在那儿度过童年。

1985年　在纽卡斯尔理工学院(今诺森比亚大学)学习设计与艺术。

1989年　成为以伦敦为基地、做设计咨询的丹吉尔公司的合伙人，在这里他的工作范围很宽，从电动工具到洗涤池。

1992年　搬到旧金山，加入苹果设计团队。

1998年　被任命为苹果公司工业设计部的副总裁。最初的iMac投放市场，在第一年里就卖掉了200万台。

1999年　推介苹果的iBook，22英寸的影院显示器，PowerMac G4 Tower和iSub。

2000年　苹果的G4 Cube投放市场。

2001年　苹果推介笔记本电脑Titanium PowerBook和iPod便携式MP3播放器。

2002年　受葵花启发产生的、带15和17英寸铰接式显示器的iMac投放市场。开始生产eMac——iMac的一款专门为教育部门使用的款式。

2003年　苹果的12和17英寸笔记本电脑PowerBook投放市场，其中17英寸的厚仅1英寸，重仅6.8磅，是世界上最轻薄的17英寸笔记本电脑。成为赢得设计博物馆年度大奖的第一个设计师。

2004年　多色彩的iPod mini和超薄的iMac G5投放市场。

2005年　被任命为苹果设计的资深副总裁。Mac Mini投放市场。

“那是最小的、最简单的实用造型，那也是真正的简朴”，他说：“它看起来简单，因为它确实如此”。

第13课

塑料技术公司：土豆削皮刀

塑料技术公司研发创造的土豆削皮刀，是产品企划中值得一提的实例。该公司是一家小公司，仅有170名员工，它制造范围广泛的塑料家用小商品，绝大多数是厨房用品。开发土豆削皮刀以前的年销售收入为800万英镑，年利润为90万英镑。公司管理层决定通过设计，将企业转向生产高附加值的产品。

他们的产品企划过程为：决定开发战略(保持他们商业上的价值/资金比，但不生产市场上的奢侈品)→分析450多种不同的竞争产品→进行定性和定量的市场调研，以发现消费者要求和需求→发现值得开发的新产品种类→对现有43种土豆削皮刀的设计特性进行分析→修改土豆削皮刀→财务可行性分析、成本和投资回收期计算→判定机会合理性→开发新产品→塑料技术公司新的土豆削皮刀。

	研发	创新设计	上市时间	生产工程	技术市场	专利
先驱型	×××	×××	××	××	×××	×××
响应型	×	×××	×××	××		×
传统型					×××	
依赖型					×××	

提议的产品：土豆削皮刀		设计者：约翰	
信息来源：蓝(市场部)		日期：1999年元月15日	
产品要求	要求或愿望	要求类型	基本/性能/激励
必须看上去卫生	要求	市场	基本
必须握起来手感舒服	要求	市场	性能
应极锋利	愿望	市场	激励

提议的产品：土豆削皮刀		设计：约翰
1999年元月15日		
性能要求	设计要求	设计规范
必须能从土豆上挖去凹眼	必须有圆凿子。 必须设计成不需紧握手把就能使用此圆凿子	圆凿子必须靠近手把。 圆凿子应能在不改变削皮刀上的手把时使用

第14课

技术侵蚀它自己

设计师们在他们设计的产品功能的技术细节上投入了许多精力。世界还不为我们所知，它是那么复杂，问题可能很容易就冒出来，特别当你的产品必须与已经建立的环境的其他部分发生关系时。

最令人恐惧的实例之一(从一个社团的立场上看)是目前发现由克利普托尼特公司制造的最普通的自行车锁可以用Bic牌圆珠笔在大约10秒钟之内打开。该想法认为，这种圆柱式锁的圆形开口在尺寸上跟最普通的Bic圆珠笔的圆柱相同，而圆珠笔的塑料可以向下斜插，这样圆珠笔芯就可以自动就位。有人放映了关于这种技艺的VCD，几天之内骑车人都知道锁自行车是没用的。只经过了几天，牢固得足够抵御地球上最有经验的自行车窃贼的一种技术却可能被一个孩子用一支笔就对付了。

有趣的是，克利普托尼特公司经过多年努力想要制成一种经久耐用的锁，而且除了液氮或乙炔火炬之外能够抵御任何东西，像钢锯、撬棒、汽车千斤顶。然而，一个完全无关领域里的一件东西，竟被发现与他们的自行车锁有着灾难性的“相容性”。

世界上投放的产品越多，产品Q与产品X发生不可预期的问题的机会也越多。在自然世界里有某些相似物，例如植物生成的许许多多化学物质用于抵御其他植物制造的大量化学物质等。谈到化学物品，决不要将氨类产品与漂白剂混合——那将产生有毒气体，比原来的两种产品更坏。这又是一个出乎我们意料的失败。

回到机械世界，另一个还没有被克服的灾难性相互作用是SUV(Sports Utility Vehicle的缩写，即运动型多功能车——编者注)车缓冲器的高度问题。SUV汽车上的大多数缓冲器(这是在缓慢变化的)要比小轿车或微型面包车的侧梁高。当更小的轿车被SUV车侧面冲撞时，SUV就会骑上小轿车的内部结构并侵入乘坐区域，这增加了乘客伤亡的几率。我有一张图证，是当我乘坐的汽车被福特“探险家”(约是1995年生产)的驾驶员侧面撞击时的情况。驾驶员的受伤程度要比两辆同样高度的车相撞时糟得多。看到的事故车表明，SUV车显然完全避开了小轿车的底梁，而使小轿车左侧乘坐区变形超过1.5英尺。缓冲器位置较低应可减小这种伤害。可能吧！

我们这儿并不是要去研究粗心大意或者过错的问题，但是某些此类例子表明，即使一个产品在其应用范围内是完美的，其外部也还有另一个可能使事情失常的客观世界。

第15课

右脑的复仇

比较准确和符合逻辑的说，左脑思维带给我们信息时代。现来临的是概念时代，由艺术气质、移情作用和感情支配的时代。

20世纪70年代中期当我还是个在美国中部中产阶级家庭里成长的小孩时，父母们端出的常见忠告是：获得好成绩，进入大学，谋求一份工作，以获得体面的生活水平和社会声誉。如果你的数学和理科很出色，那么成为一名医生吧！如果你擅长语言和历史，那么就做一个律师。假如你稍微迟钝，口头技能又需改进，就成为一名会计。后来当人们的办公桌上出现了电脑，而CEO(首席执行官)们出现在杂志的封面上时，那些数学和自然科学学科好的年轻人选择了高科技行业，同时其他人成群结队出现在商学院里，认为成功就是MBA(工商管理硕士)。

税务代理、放射科医师、金融分析师、软件工程师，管理学的领袖彼得·德鲁克给这些职业一个稍微有点虚幻但不朽的名字：知识工人。他写道："这些人把他们在学校里学到的知识用到工作中并获得报酬，而不是靠体力和手工技能"。这些人中成为佼佼者和使他们能获得社会最高奖赏的，是他们的"获知和应用理论及分析知识的能力"。我们中的任何人都能加入到这个行列中。只要我们努力学习并遵守精英阶层政体的规则。这就是专业成功和个人价值实现的道路。

当我们未曾休息片刻，有趣的事情发生了：这个世道变了，未来不再属于那些像计算机一样条理清晰、快速和准确的人，未来属于另一类有着不同思维方式的人群。今天，——在经济从繁荣到萧条的不确定之中——是个解释到底发生了什么的隐喻。这就是我们头脑中的公理。

科学家们早就发现，一条神经学上的梅森-狄克逊线把我们的大脑分成两个区域——左右两个半球。在过去的十年里，部分得益于功能性磁共振成像的进步，已经开展了更加精确地确认大脑两半边的职责分工的研究。左脑掌管顺序、文字和分析工作。而右脑关注内容、情感表达和综合。当然，人类大脑有着1000亿个脑细胞以及更多数量的神经链接，是不可想象的复杂。两边大脑协调工作，我们做每一件事情都要同时调动两边的大脑。但是，我们大脑的结构却可以帮助我们揭示这个时代的特征。

直到近来，那些在学校、工作和商业上成功的能力都表示了左脑的特征，由SATs测得、由CPAs开发的左脑的特征，直线的、富有逻辑和分析的天赋。今天，这些能力仍然必需的，但不再够用了。在这个被外部资源充斥、数据泛滥、选择令人窒息的世界里，这种能力在精神上更接近于右脑的特征——富有艺术感，移情作用，观看一幅大图画，追寻超常力。

在过去五年的精神紧张的骚动中，一股缓慢而强烈的变革出现了。我们充分准备了的信息时代即将结束，取而代之我们称之为的概念时代，在这个时代里，能否掌握那些我们几乎忽视和低估的能力，必将成为区分我们领先和落后的分界线。

对你们中的有些人，这个变革——从信息时代建立在逻辑、顺序能力之上的经济转换到概念时代建立在创新、移情作用能力之上的经济——听起来令人愉快。“你让我应接不暇！”我能听到画家和护士们欣喜若狂。但对另一些人而言，这简直是胡说八道。“证明它！”我听到程序员和律师们这样苛求。

好！为了使你信服，我将用机械论的因果语言解释这种变革的理由。

……

第16课

著名设计师：路依吉·柯拉尼，迈克尔·格雷夫斯，飞利浦·斯达科

路依吉·柯拉尼

煽风者，挑战者，审美家，设计领袖，实际形状和完美形态的创造者，圆滑波形样式的倡导者，人机工程学、风格和轮廓的教授，哲人宝石的托管者。

地球是圆的，天空所有的物体也是圆的，它们都在圆形或者椭圆形的轨道上运行。

“我只是个自然的解译者”，柯拉尼说。

“自然界里的事物也是由曲线构成的。”

这种彼此环绕旋转的球形世界的相同图形一直追随我们直至微观世界。甚至在与性元素相关的繁衍方面，我们都被圆形所包围。“我为什么还要加入到想让每个东西都有棱角的迷路人群中去呢？我正在继续伽利略的哲学：我的世界也是圆的。”

迈克尔·格雷夫斯

从1962年起他就在普林斯顿教授建筑学。他为阿莱西公司创造了最畅销的9093壶。

“当1980年元月初我们首次拜访他时，迈克尔·格雷夫斯告诉我们，此后他至少要花一半时间在设计上了。

这是个决定性的声明，与迈克尔·格雷夫斯在这个领域的巨大潜力相符合。他不喜欢理论，尽管有一次公开承认他希望开发‘美国式设计’。

20世纪80年代到90年代间，在任何情况下他都显示出了一种难以置信的能力来亲近一般大众的口味。他显现出他能对公众施以魔法，就像与我们合作的很少几个设计师那样。”

飞利浦·斯达科

他1949年生于巴黎，是我们时代最新颖、最有创造性的设计师之一。尽管得到许多重要奖

项的认可，他却自认为是个“日本建筑师、美国艺术主管、德国工业设计师、法国艺术指导、意大利家具设计师。”

“斯达科1986年在一个法国设计项目上开始同我一起工作。

他是个我梦想中的鲜活例子：设计，真正的设计，面向制造和贸易世界而永远充满高度的创造，带来的结果无需单独在技术或者资产平衡表的层面上证明。

一项真正的设计工作必须调动人，传达感觉，唤起记忆，令人惊异，超越界限……总之，应该是诗歌般的。设计是当代最倾向于诗歌般的表达形式之一。

而我知道，这个伟大的幻想家仍然藏有大量令人惊奇的秘密武器，尽管他以退休相威胁！”

第17课

用于评估理念的DE-BONO六顶思维帽子

白色帽子：信息
- 需要什么信息？
- 可以得到什么信息？
- 遗漏了什么信息？
- 如何获得我们需要的信息？

黑色帽子：逻辑否定
- 警告
- 风险分析
- 临界性(危险程度)

黄色帽子：逻辑肯定
- 利益
- 价值
- 价值敏感度
- 如何能很好工作？

绿色帽子：创造性
- 新理念
- 其他可供选用的理念
- 可能性
- 刺激——横向思维

红色帽子：将感觉向前推进
- 感觉
- 直觉
- 情感

蓝色帽子：思维过程
- 如何定义解决办法？
- 决策

● 目的是什么？

De-Bono的6顶思维帽子的作用：

通过把每个人的想法平行地结合在不同的思维模式中， 将潜在的矛盾冲突转化为合作， 区分不同的思维评价模式， 形成建设性的决策， 对理念作出评估。

虚拟的设计

如果你还记得几周前的事，我们曾聊过一会儿通过设计把虚拟的信息整合到实际世界中去这种机会越来越多。但是有趣的是，随着目前网上的多玩家游戏的大量涌现，反对声也一直在激增。作为一种出路，设计师正在转向虚拟世界，有时候甚至作为他们设计的收入来源。将会惊喜不断吗？艾米·韦博是个服装设计师。她的风格很快就会出现在纽约街道，就像20世纪80年代中期麦当娜的全套用具那样。以下是最近一次会面时关于她的设计过程的谈话：

“不错，通常我在街上留心好的设计素材作为开头。我在皮包里总放一台数码相机，看到什么酷了就对准它，像建筑物的前立面、垃圾桶、我喜欢的衣服式样。我就收集所有这些特征。我把它们都输入到我的计算机里。然后当我看到一个我喜欢的式样时，可能就是在地铁上看到的一件小女孩很酷的裙子，有时是在回家路上看到的某些别致的东西。”

但是为了使你熟悉她的设计，你要开始接触虚拟的东西。她的衣服和服饰只能穿过她的专卖小商店、在一种多玩家的在线游戏“第二生命”里得到。她不孤独。“第二生命”的使用者数量在不断增加，这些人关注于设计精致的家、汽车、家具，甚至是SciFi(科幻)型的车辆。

有些项目的精心制作使它们几乎成为“第二职业”。芭克拉·罗兹，一个现实生活活动的组织者最近策划了两个游戏朋友的在线婚礼。这件精心安排的事情发生在一艘船上，需要几个星期的准备及几百美元真钱。而更大的制作是为了先要把鳄鱼岛改造成莱曼·弗兰克·鲍姆的“绿野仙踪”里的仙境。那儿甚至还有带空中烟雾文字的女巫。

当然，那不都是玩笑和游戏。杰西战争是一场可笑的血腥冲突，这场冲突使其模拟器的许多“玩家-杀人者”部分之一在短时间爆发的。这场冲突最终烧尽他们自己，连同玩家构筑的围墙和枪炮，但是又有谁知道将发生什么样的暴力呢？另一种有趣、但是令人不安的使用游戏是“白屋子”，一种在“最大痛苦游戏引擎”中进行的光射击。艺术家创造了使人想起以前暴力的场景，这种暴力使观看者把历史融入这种场景。

目前，所有这些都是琐事——很有趣，但不是真实的设计，是吗？好嘛，在未来几年里我们的时间将会更多地花费在网络上，而且大概会用虚拟手法来表现我们自己或其他人，那时我们的真实设计技巧会不会与他们现在做的一样起作用？当计算机越来越成为一种真实的模拟时，越来越少的二维抽象、真实的设计、人类工程学及使用者的理解会变得越来越重要，就像他们今天在其真实世界的相似物一样。假如你有兴趣，“第二生命”可能需要酷的产品。那你还等什么呢！

第18课

塔珀家用塑料制品公司的“FlatOut!”家用塑料储物箱

塔珀公司的“FlatOut!”家用塑料储物箱是为储物时需要灵活性的消费者、刚离开学校工作的职业人士和露营者设计的。它的折叠设计使“FlatOut!”能按照储藏的食品或节省空间的堆叠来调整适当的容量。该容器关上时可达到完全水密。清洗时，可以像盘子一样叠起来放进洗碗机洗涤。2004年12月，该产品被优良家务管理协会授予“好买卖奖”，并被“财富”杂志在其2004年12月号里命名为年度25种产品之一。新近，它又由德国消费者塑料制品协会授予“2005年度产品”奖。

“林地(Timberland)”牌旅行装备

“林地”公司开发了一种新概念的装备，以解决旅行者在决定要往行李里装几双什么样的鞋时总是面临进退两难的问题。通常的情况是，他们所带的鞋可能并不能满足他们将要进行的各种活动的需要，而他们带的鞋却要占据宝贵的行李空间。“林地”牌的这种旅行装备是一种革命性的模块式的鞋类。通过将鞋的外观(鞋面)与核心功能(鞋底)分离，并使它们可以互换，穿用者可以同时拥有多种样式及功能的选择，以满足对鞋类一系列的需要，从远足到商业会谈，从外出就餐到阴雨天气。三双鞋面，加上两种鞋底及一双防水袜，通过各种组合就相当于12双普通鞋。另外，这些部件压扁了之后，只占最普通鞋的一半空间。

关于人体工程学

大多数产品是给人们以某种方式使用而设计的。在仔细审视时，即使对于最简单的产品，产品与使用者间的界面也往往是很复杂的，且很少被充分理解。因此，产品设计的这一面，常常提供了概念设计方面丰富的灵感来源。任务分析就是要通过观察和分析来探究产品及其使用者之间的交互关系，并且利用其结果来产生新的产品概念。它给设计师提供了消费者如何实际使用该产品的第一手体验。通过这个过程，能激励概念的产生以便改进使用者界面，并为随后应用人类工程学和人体测量学的设计方法铺平道路。任务分析包括了产品开发的两个非常重要且又高度专业化的方面：人类工程学和人体测量学。

英文ergonomic(人类工程学)这个词，起源于希腊字“ergon”，意义是工作，因此它的字面意思是对工作的研究。早期的人类工程学研究确实是研究在其工作环境中的人，但现在它的使用范围变得宽泛得多，已经涉及人和人工制品之间的互动。人类工程学就其学科名称而言就是一个研究课题，它涵盖了解剖学、生理学和心理学，并已很好地应用于设计上。就大多数产品设计师而言，应付人类工程学的最好方式就是以“须知”为基础。如果你着手搞的设计项目涉及与产品的某种形式互动，则你再去进行该任务目前的人类工程学知识研究。当然，对大多数设计工作而言，对人与产品间互动关系的足够认识，可以通过对人执行相关任务时的情况进行观察并由此得到对有关项目的第一手了解而达到。

人体测量学则是人的尺寸测量。在设计为人所使用的产品时，不可避免地要以人体测量作为产品的尺度基础。就人体测量学而言，通常找到数据并不是问题。问题在于如何使用这些数据。

第19课

关于飞利浦设计公司

1998年初，飞利浦设计成为集团内的一个独立公司，不仅可以服务于飞利浦集团内部的客户，也可以服务于集团外的客户。

飞利浦设计公司认为该公司的任务是：通过提供有竞争力的、卓越的高端设计方法，为消费者、公司股东以及社会整体创造价值；超越消费者的预期，促进技术的融合，满足制造需要，致力于提升生活质量及拥有和使用的方便和愉悦。

它的观点是，创造关注人的成长的设计，以使人们能够彼此和谐并与自然和人造环境和谐地生活。“通过重视人的价值来达到为人创造价值。”

这种设计方法是基于这样一个前提的：除非设计以研究为基础并以人为中心，否则设计决不可能一直成功。这就是为什么该公司雇佣了这么多人类科学方面专家的原因。这就是为什么该公司常常与外面的研究单位和合作者一起联合进行许多设计研究项目的原因。这就是为什么该公司要采用一些被认为是超出了“设计”边界之外的方法的原因。这就是为什么公司能够通过从概念和策略到产品实现和制造的不同层级的商业创造过程来回应其客户的需要的原因。

设计师斯蒂芬·迪厄兹与他的一些设计

长沙发“库奇(couch)”(见右图)是由蜂窝状纤维素材料构架并填充聚苯乙烯小球做成的。这些聚苯乙烯小球在运抵目的地国家后才被放入。因此，沙发的构架可以折叠起来运输以便节约运输成本。

“忒玛(tema)”的灵感更多来自于现代服务设计。因为在今天相对于大份量的面条和超大号的美食盘，传统的刀叉小得都要看不见了。在这个问题上，斯蒂芬·迪厄兹证明了他是考虑这方面问题的先驱。他的设计结果是一套更大的经典式扁平刀叉。

“大箱子(big bin)”是一个多用途的储藏系统，由可叠加的ABS塑料容器组成，可用于储藏文件、玩具、洗衣烘干房物品、周转产品及运输各种物品。通过像建筑模块那样的组合，各个部件可以快速地建成一个稳定的储物架；每个容器的侧面都有把手，这个把手可以用来作为垂直和水平构架时的互锁零件。

当使用“洁尼厄(genio)”时，食品是从厨房直接端到餐桌上的。这一系列的不锈钢锅只是简单地加上了一个瓷的“壳”，这样就使食物能够保温更长时间。

斯蒂芬·迪厄兹1971年生于慕尼黑市的弗赖星。在完成了他的木工训练后，他在印度的孟买和浦那为一个家具制造商工作了一年，随后在斯图加特美术学院学习工业设计。作为助理设计师，他在美国为理查德·萨珀公司工作了一小段时间；并为加拿大的康斯坦汀·格尔希克公司工作了大约两年；之后在2002年，在慕尼黑开设了自己的设计工作室。

第20课

让技术变得温暖又模糊

高技术消费者正在与手工艺人形成一个靠不住的联盟，他们追求美观愉悦的电子产品配件，而拒绝大众产品。

对于个人装备，电子产品迷们追求的是个性，而不仅仅是苹果或其他一些公司可以提供的屈指可数的几个艳丽的颜色。因此，一大批手工制作的产品进入了网上市场，例如带有绣上去的心形和星形的手工缝制的绒布，绣有黄色小鸟和蜜蜂的垂花彩，以及钩针编织的保暖套。

就拿瓦内萨·布拉蒂为例来说吧，她在去年购买了iPod之后，试图寻找一个套子，既能保护她的这个小玩意，又可以展示她对时尚的敏感。由于对市场上的大批量生产的产品不满意，她放弃了这些主流产品，而试着自己动手制作一个既令自己满意也最终令他人在审美和保护质量上满意的iPod套子。

布拉蒂，27岁，是“大丁草”设计公司的创建者和拥有者。她住在得克萨斯州的埃尔·帕索市，她在她自己的网站上，以及像“像素女孩”商店这样的地方，以30～35美元的价格出售她制作的iPod外套。

这个趋势证明，手工技艺和高科技技能不是相互排斥的个性特质。因为对于大多数像布拉蒂那样的手工制作者来说，他们使用他们自己的产品，因此消费者可以得到保证，一方面该产品是手工制作的，能够确信产品既是手工制作的，另一方面它又可以保护他们的iPod或是膝上电脑。

从西雅图到布鲁克林，手工制作者们制作着一些价格合理的物品，如电子产品的包和保护套。由于这些产品都是手工制作的，因此供应量往往有限，而不像那些大批量生产的产品。这些产品没有两件是完全一样的。

一份主张自己动手做的杂志《准备制造》的主编肖莎娜·伯杰认为，这并不是一次新的手工业革命的一部分，而是手工业者适应新出现的技术的结果。

伯杰说：“我的想法是，当茶壶出现的时候，人们就开始制作茶壶保护套，而现在，人们给iPod做保护套。”

消费者们看起来很欣赏这种高技术和手工装备的配对，不论工艺是否真是这种结合，对这些手工制品有兴趣购买的人是些二三十岁的小玩意迷，伯杰说，其中一些人买这些手工制品是因为电子产品的更新换代太快了。

她说："我们规划了将这些电子产品逐步过时，像你的 iPod会在一两年内被淘汰。从我们读者的热度为基础来看，人们已经警觉到了这一点，并且他们对于在生活中拥有一个手工制作产品的主意表示赞同。"

根据工艺品组织发展协会的一项调查，2000年全美国估计有12万6000多人部分或全部收入来自于销售手工制作的产品。

全国工艺品协会会员部主管安·巴伯说，大约70%的职业手工业者在他们自己的网站上销售产品，而且这个数量还在增长。

28岁的珍尼斯·赫德列认为，这些产品让消费者能够将本来死板的或黑或白的电子产品个性化。赫德列在她运行的"卓越"里出售她自己及其他人手工制作的物品。住在西雅图的赫德列以34美元的价格出售装饰有录音机和动物图案的绒布CD包。

这个包，还有其他她使用的东西——一个钩针编织的手机袋，和一个粉、白两色的手工制作的膝上电脑包——都将科技个性化了。这种个性化，对于消费者和制作者都很有吸引力。

"人们希望有多种选择，人们希望个性化的东西。我认为正因为电子产品公司没有真正提供这些东西，因此大家只好自力更生。如果你买不到一台粉红色的膝上电脑，那就做一个粉红色的包吧"，赫德列说。

罗布·凯林是6月份开张的网上手工制品市场"埃特喜"的共同创始人，这个网站现在有1700个注册了的手工制品制作者在出售产品。

他说："直到最近，由于是手工制作，才使得这些产品形成了一个专业的市场。而在过去，(手工制品市场)曾经是唯一的市场。"

最后要说的是，虽然如此，不管是高科技还是低科技，可能所有这些归根到底只是电子迷时髦的因素，27岁的帕特丽夏·瓦尔迪兹说。她得到很多对"大丁草" 设计公司机器人式样的iPod套的评论。

"我想人们能注意到它真是非常的酷"，她说："这个东西与众不同。"

第21课

自灌溉花盆

这款自灌溉花盆可以保证植物爱好者的植物不会因为他们外出度假或者是忙碌而干枯死亡。它的设计不但美观而且实用。在陶瓷花盆的漏水孔中装有虹吸芯。植物栽种在花盆中后，花盆再放置在一个盛了水的玻璃容器的顶上。虹吸芯垂在盛水的容器中，植物通过虹吸芯吸取所需的水。平均来说，这个玻璃容器可以装一星期的用水。这个设计让使用者能够得到简单直观的反馈。只要对水位看上一眼，就能告诉你是否该加水了。

"迷你"运动表

这款"迷你"运动表的特色是，它的数字显示从水平方向转变为垂直方向，这样当使用者在骑车、步行或驾驶汽车时，他们的手腕与视线处于一条直线上，这种人机工程学的考虑可以让使用者很容易读取时间。由于使用了合成橡胶包裹的弹簧钢芯的软表带，使它可以适应不同的手腕大小，易带易脱，并且舒适。此款手表在高档商品专卖店，如"芭妮"、"航班001"、"慕玛"和"莎芙慕玛"，及一系列小商品零售商及目录上销售，它是纽约"慕玛"目录里最畅销的产品之一。第一轮产品已销售一空，第二轮产品现在正在生产。由于此产品非常成功，"迷你"公司计划在2005年投放更多样式和型号的"迷你"运动表。

配备"宝"衬里的K2 T1靴

此项目的目的是要重新设计K2滑雪板公司最高档的T1滑雪靴，将新技术与新式样融合到设计中去。传统上滑雪板靴有两个束紧系统，衬里一个，外壳一个。T1靴有一层称为"宝-阿克赛斯"(Bao Access)的衬里，可以使滑雪者不用脱掉外层就能细调衬里的松紧。要调紧时，只需在揿下的状态转动位于靴筒上部的称为"宝"的转盘。如果需要立即解除这套鞋带的束缚，使用者只需将"宝"转盘弹出。作为2004～2005年度K2滑雪靴系列的旗舰型号，T1带动销售比上一季度高出了25%。

设计的滋味

完美和激情：出于这两个理念，阿莱西公司和"生物多样性慢餐协会"彼此挑中了对方，他们联合起来推广一个重要的创新，其特点是将美食的享受和设计的创造性结合起来。

所有由当地小范围生产的番红花、盐、小扁豆、大米、蜂蜜、咖啡、香草、杏仁、无花果和玉米饼等10种食品产品，它们是从270多种中挑选出来的、受意大利与国际慢食基金会主席保护的，各与一个与其风格相应的阿莱西公司出产的用于盛放和上餐的用品一同出售；这种风格可以提供"附加价值"。整个包装细致简洁却幽雅自然，盒内还有一本介绍该生物多样性项目的历史、目的和物流的小册子。

生物多样性慢餐协会在选择合作伙伴时一贯小心挑剔，它追求"做事"的激情与"把事情做好"的文化之间的紧密结合，这与阿莱西公司的信条是一致的。

通过圣诞节期间在其零售店发售这首批10种食品，阿莱西公司与这些小食品的生产者充分合作并积极支持他们，而这些生产者直接地、没有任何有经济关系的中间环节供应他们的食品，同时，这些生产者通过这种合作可以依赖一个重要而优质的市场，向欣赏美味和品质的大众开放。

第22课

消费者需求与新产品开发

消费者需求、竞争对手的对策以及安全和环境立法朝着更严格方向的改变等因素迫使企业不断地开发新产品。通过快速引入新产品、有意识地缩短产品的市场寿命这样一种管理战略，是对付动作迟缓的竞争者的战略武器。

新产品的开发可以分成三大类：改进型、整体开发型和技术创新型。

开发新产品十分重要，又充满了风险。为什么它的风险如此之大呢？是什么让许多产品归于失败呢？因为在产品开发初期，不确定性十分高。你不知道尚待开发的这个产品是什么样子，怎样做出来，成本有多高，购买者对它有什么想法；总之，新产品开发是个多因素的问题，许许多多的因素都会决定产品的成败(例如如何引起购买者注意，零售商的接受程度，工程的可行性，产品的耐久性和可靠性等)。

概括而言，以下三大方面中的诸多因素使新产品开发的成败有了最重要的差别：

(1) 市场导向：最主要、也许最明显决定经营成功的因素是市场的区分和客户的价值观。如果客户认为该产品比其他竞争产品好得多，价值高，则其成功率要比差别不大的产品高出5.3倍。另一些对市场导向起作用的因素是让产品尽早面市和高效率地向市场推出产品的能力。

(2) 早期的可行性评估和规范：这方面有两个重要因素。着手开发新产品之前，对产品作过充分和严格的可行性评估的，其成功率要比那些不作评估的高2.4倍。清晰和确切地确定产品的设计规范，其成功率会提高3.3倍。

(3) 新产品开发过程的质量和开发团队：一般来说，技术活动始终是以高质量进行的，开发产品的成功率会高2.5倍。尤其是，公司的技术技能如果能适应新产品开发所需开展的各种活动时，成功的机会会增加2.8倍；公司的销售和市场开拓能力能适应新产品推广需要的，成功机会会增加2.3倍；公司内部技术人员和营销人员之间“工作和谐”程度高的，可使新产品成功的几率比“很不和谐”的环境时高出2.7倍。

为了使新产品的开发获得成功，通常可采用风险管理漏斗的方法来控制风险。风险管理漏斗实际上是关于新产品开发中新产品的风险和不确定性变化的一种思维方法，代表了新产品开发中所作的决策。

第23课

扼杀思想理念的21种情形

(1) 忽视；

(2) 有意躲避；

(3) 不屑一顾；

(4) 嘲笑；

(5) 过度赞扬；

(6) 谈及从未这样试过；

(7) 证明它不是新东西；

(8) 认为不符合公司政策；

(9) 提及其成本花费；

(10) 过去已经试过了；

(11) 鼓励创造者寻找更好的理念想法；

(12) 找个有竞争力的理念(想法)；

(13) 提出一大堆它不能工作的理由；

(14) 修改得面目全非；

(15) 怂恿对拥有权提出质疑；

(16) 把其他理念联合起来指责它；

(17) 试图将它逐渐地瓦解掉；

(18) 对理念发明者进行人身攻击；

(19) 给予技术性打击；

(20) 推迟实施；

(21) 听任一个委员会去审理该理念想法。

索尼公司和东芝公司的随身听

20世纪70年代，随身听首次做出样机的时候，市场上已有带式录音机作为录音和放音的产品。而随身听非但没有录音功能，而且只能一个人通过耳机使用。索尼公司市场部认为随身听是一个“愚蠢的产品”。但索尼公司的主席盛田先生本人是个高尔夫球迷，他的梦想就是能一边打高尔夫球一边听音乐。他亲自进行了试用，十分喜欢。他的意见否决了市场部的怀疑态度，这个产品获得了巨大的成功。在随身听的市场需求调查中，成功的调研正是得益于确定了正确的问题：“出去散步或活动时你想听音乐吗？”

而东芝公司在开发随身听产品时，分析了在市场上索尼公司的产品，它们诉求重点是“高品质”、“高技术”。因此东芝将产品定位为“高时尚”，目标定位于年青人，尤其是少女消费层，结果在竞争中也大获全胜。

第24课

关于交互设计的一些知识

社会交互设计

社会交互设计(缩写SxD)正在浮现，因为我们许多计算装置已经联网工作而开始信息能力的一体化。电话、数字辅助设备，以及从计算机到游戏机无数联网的装置推动了会话和社会互动。社会交互设计说明了使用者间以及使用者与他们的装置之间存在的互动。人与人之间信息沟通、讲话和书信的原动力，谈话和互动的语用论(研究语言符号与使用者关系的一种理论——译注)——现在在社会技术使用中，这些都成为了决定性的因素。而这些因素在由认知科学采用的理性选择方法中要比在社会学、心理学和人类学中要描述得少。

情感交互设计

在整个交互设计过程中，设计师们必须意识到在他们设计中影响目标使用者情感反应的关键方面。对于产品传达正面情感以及避免负面情感的需要对于产品的成功是具决定性的。这些方面可以提出几个，包含正面的、负面的、有关动机的、学习的、创造性的、社会的和劝诱的影响等。能够帮助传递这些信息的一个方法是使用表达界面。例如在软件中，使用动态图标、动画和声音能够帮助沟通操作状态，创造一个互相作用和反馈的感觉。界面方面诸如字体、颜色托板、绘制图案也能影响一个界面的感知效力。研究表明，情感表达能够影响使用者的可用性感知。

情感和愉悦学说解释了人们对交互式产品使用的反响。这包括堂·诺曼的情感设计模型，帕特里克·约旦的愉悦模型，以及麦卡锡和赖特的体验构架技术。

交互设计范畴

交互设计师们工作在许多领域，包括软件界面、(商业)信息系统、因特网、物质产品、环境、服务，以及可以是这几方面结合的系统。每个领域要求各自的专业技能和方法，但是存在对所有交互设计共同的方面。

交互设计师们经常以多学科混合团队的方式一起工作，因为他们的工作要求在许多不同的范畴有专门技术，这包括图形设计、编程、心理学、使用者测试、产品设计等(更多的相关领域参见下面)。因此，他们需要对这些领域有足够的了解以便同专家们一起有效工作。

历史

交互设计这个术语是由比尔·默格里奇以及比尔·佛泊兰克在20世纪80年代后期第一次提出的。对于佛泊兰克来说，这是计算机科学术语“使用者界面设计”转到工业设计职业的一个改编。而对默格里奇，是在1984年对包含软件的产品绘制一个图标，查阅工业设计的应用时对“软外观”的一次改进。

1989年，吉莉安·克拉泊通-史密斯在伦敦皇家艺术学院建立了交互设计硕士点(原来授权为“计算机相关设计”，而现在是“设计交互”)。2001年，她帮助在意大利北部创办了依伏里安交互设计学院，一个单独专注于交互设计的小学院；2005年10月该学院迁移到了米兰，与朵玛斯学院合并。今天，交互设计已在世界范围的许多学校里教学。

附录II：全书词汇和短语索引

A

a succession of　一连串　【17】

accessible design　接近性设计，亲近设计　【21】

accuracy　精确度，正确度　【24】

adage　格言，谚语　【9】

adhesive　粘合剂　【12】

adoptive design　适应性设计　【21】

aeon　万古，永世，十亿年，漫长岁月　【16】

aerodynamics　空气动力学　【16】

aesthetic appeal　美学诉求　【2】

aesthetics　美学，审美学　【22】

alloy　合金　【8】

analogy　类推，模拟　【5】

analysis　*n.*　分析，分解　【5】

anthropologist　人类学家　【23】

anthropology　人类学　【8】

appeal　*n.*　请求，呼吁，吸引力；要求

vi.　求助，诉请，要求　【2】

arm with　用……武装　【6】

aspect　面貌，(问题的)方面　【10】

assistive design　帮助设计　【21】

associated with　与……有关联的　【10】

attentive mode　注意模式　【3】

attract　*vt.*　吸引　【1】

　　vi.　有吸引力，引起注意　【1】

attractiveness　魅力　【1】

audience　观众，听众　【10】

authority　权威，威信，权力　【8】

(be)aware of　知道，察觉到，意识到　【24】

B

barrier-free design，design without barriers　无障碍设计　【21】

basic function　基本功能　【18】

battery charge　电池充电　【10】

beige　米色，浅褐色的　【12】

benefit　*n.*　利益，好处

　　vt.　有益于，有助于

　　vi.　受益　【6】

big seller　大卖家；畅销品　【8】

bio-dynamics　生物动力学　【16】

boundary　边界，分界线　【6】

brainstorming　头脑风暴(法)　【5】

brainwriting　创意激荡(法)　【5】

brand　*n.*　商标，牌子；烙印

　　vt.　打火印；侮辱　【10】

break-even point　盈亏平衡点　【13】

bug catching　缺陷捕捉　【24】

business opportunity　商业机会　【13】

business strategy　经营策略　【17】

business success　商业成功　【15】

C

case　案例　【10】

catamaran 双体船 【16】

catering 公共饮食业，给养 【8】

champion 冠军；支持，拥护 【8】

channel 通道，通路 【9】

chrome 铬，铬合金；镀铬 【8】

clairvoyance 洞察力，千里眼 【9】

cliché (法语)谚语，陈词滥调 【5】

client 客户，委托人 【23】

clinical 临床的，病房用的 【10】

close-knit 组织严密的 【12】

collect *v.* 收集，聚集，集中，搜集 【5】

collective notebook 集体笔记本(法) 【5】

come up with 提出 【9】

commission 委托；任命 【11】

commit(to...) 献身，效忠 【12】

common sense 常识 【20】

compatibility (with...) *n.* [计] 兼容性 【9】

compete *vi.* 比赛，竞争 【14】

competing product 竞争产品 【13】

competing product survey 竞争产品调研 【14】

competitive advantage 竞争优势 【20】

competitive weapon 竞争武器 【20】

comply with 遵循，遵照 【4】

conceive 构思，设想 【8】

concept 观念，概念 【6】

concept and idea selection 概念和方案选择 【7】

concept design 概念设计 【2】

concept selection matrix 概念选择矩阵 【7】

conceptualize 使有概念，概念形成 【8】

(be)conducted to 旨在 【15】

conduction 传导性 【15】

conflict　矛盾，冲突，抵触　　【22】

conformance with　与……的一致性　　【9】

conquer　征服，占据　　【8】

consequent　作为结果的，随之发生的　　【12】

consultancy　顾问，咨询(工作)　　【12】

context　上下文关系，文脉；背景　　【12】

continuation　连续性　　【4】

contour　轮廓，等高线　　【16】

controversial　有争议的　　【16】

coordinate　*n.*　同等物，坐标(用复数)

adj.　同等的，并列的

vt.　调整，整理　　【8】

core　果核，中心，核心　　【6】

core benefit proposition　核心利益主张　　【6】

create　创造，创作，引起，造成　　【5】

creative　创造性的　　【19】

creative thinking　创造性思维　　【5】

creativity　创造力，创造　　【20】

creature　人，动物，傀儡，创造物　　【19】

criterion　(批评判断的)标准，规范　　【7】

critic　批评家，评论家　　【16】

critical　评论的，鉴定的，批评的，危急的，临界的　　【16】

cultural　文化的　　【10】

customer-oriented　消费者导向的，以消费者为中心的　　【18】

D

democratic　民主的，民主主义的，民主政体的，平民的　　【21】

democratic value　民主价值　　【21】

dense　密集的；极度的　　【12】

dependent strategy　依赖型策略　　【17】

design skill　设计技巧，设计技能　【23】

design specification　设计规范　【13】

destructive　破坏(性)的　【22】

deviation　背离，偏差　【22】

diagnostic equipment　诊断设备　【10】

disability　无力，无能，残疾；[律] 无能力，无资格　【21】

discipline　训练；学科　【22】

discrimination　辨别，区别，识别力，辨别力；歧视　【21】

disengage (from...)　脱离……　【22】

dissever　割裂，分开　【22】

dissolve(into...)　解散；溶解　【22】

distinct　独特的，截然不同的　【16】

distribution　分配，分发；销售，分布状态，区分，分类，发送，发行　【9】

diverse　多变化的，不同的　【22】

diversification　变化，多样化　【22】

dominant　占优势，支配地位，有统治权　【16】

dot sticking(technique)　贴点法　【7】

drab　土褐色的，单调的　【12】

dynamics　*n.*　动力学　【16】

E

ecological　生态的，生态学的　【22】

ecology　生态学，[社会] 环境适应学，均衡系统　【22】

efficiency　效率，功效　【24】

elaborate　*adj.*　精心制作的，详细阐述的

vt.　精心制作，详细阐述

v.　详细描述　【8】

emotion　情绪，情感　【11】

empower　授权于，使能够　【21】

enable(enebling)　使能够，授权　【23】

entertainer　演艺家，艺员　【16】

entertainment　*n.*　款待，娱乐，娱乐表演　【16】

enthusiasm　狂热，热心，积极性　【10】

entice　诱惑，诱使　【11】

environment　环境　【22】

equal　*adj.*　相等的，均等的，不相上下的

n.　对手，匹敌，同辈　【21】

equal opportunity　机会均等　【21】

equipment　*n.*　装备，设备，器材，装置；铁道车辆；

(一企业除房地产以外的)固定资产；才能　【10】

era　时代，纪元　【23】

ergonomic　人机工程学，人体工程学　【16】

ethic　伦理，道德规范　【22】

evil　罪恶，不幸，诽谤　【9】

evolutionary　进化的，渐进的　【10】

existing solution　现有方案　【6】

extent　*n.*　广度，宽度，长度；范围，程度　【17】

F

facial value　外观价值　【4】

fail to do so　未能这么做　【1】

fastener　纽扣，按钮，卡扣　【12】

feature　*n.*　面貌的一部分，特征，容貌，特色，特写

vt.　是……的特色，特写，放映

起重要作用　【5】

feel　*vt.*　摸，触，感觉，觉得

vi.　有知觉，有某种感觉

n.　感觉，觉得，触摸　【19】

fiction　虚构，小说　【8】

figure out　算出，想出，弄清楚……　【24】

finalization　定案，定稿　【23】

financial　财政的，金融的　【13】

financial model　财务模型，财务模式　【13】

first insight　第一洞察力　【20】

fit　*n.*　适合，痉挛，突然发作

adj.　合适的，恰当的，健康的

vt.　适合，安装，使适应，使合格

vi.　适合，符合　【20】

fit in with　与……配合，与……一致，适应

flatware　(刀、叉、匙等扁平)餐具　【11】

flick　轻弹，突然移动；(瞬间)照见　【1】

flick over　翻阅　【1】

font　字体，字形　【10】

forebear　祖宗，祖先　【8】

form　*n.*　形状，形态，外形，表格，形式

v.　形成，构成，排列，组成　【3】

free fall　自由落体　【15】

function analysis systematic technique(FAST)

功能分析系统技术　【18】

function tree　功能树　【18】

functional attractiveness　功能魅力　【2】

functional principle　功能原则　【6】

functional speciality　功能特点　【14】

functional value　功能价值　【2】

functionality　功能性　【16】

G

garment　衣服　【10】

genius　天才　【16】

geometric form　何形状　【4】

gestalt rules　格式塔规则　　【4】

go off　偏离　　【9】

grab　抢夺，夺取，攫取

grab one' s attention　引起注意　　【1】

ground　*n*.　地面，土地，场所，范围

　　adj.　土地的，地面上的

　　vt.　把……放在地上，打基础　　【9】

ground-rules　基本准则　　【9】

guru　宗教老师，领袖，头头　　【16】

H

harmonize(with)　协调　　【22】

harmony　协调，融洽　　【3】

head-up view　高阔的视野　　【10】

helm　舵　　【8】

heritage　遗产，传统　　【12】

hit　击中；(演出等的)成功　　【11】

hospitability design　关怀设计　　【21】

human　*n*.　人，人类

　　adj.　人的，人性的，有同情心的　　【10】

human-focused　*adj*.　关注人的，以人为本的　　【23】

humanize　赋予人性，人性化　　【10】

humbling　令人羞辱的　　【12】

hybridization　杂交，杂种培植　　【7】

I

icon　图标，肖像　　【10】

idea　想法；主意，思想，观念；概念　　【7】

idea generation procedure　概念产生程序　　【5】

idiosyncrasy　特质，特异性　　【19】

illumination 阐明 【20】

image *n.* 图像，肖像，偶像，典型

vt. 想象反映，象征 【6】

imagination 想象力 【6】

imaginative 有想象力的，虚构的 【19】

imitate 模仿，仿造 【16】

immediate appeal 直接诉求 【3】

immemorial 古老的，远古的 【8】

imperative 势在必行的，绝对必要的，强制的 【12】

imperfection 不完整性 【22】

in conjunction with... 与……联合 【23】

in harmony with... 与……协调、和谐 【23】

incubation 深思熟虑 【20】

independence 独立自主 【21】

indispensable 必不可少的 【21】

inherent attractiveness 固有的魅力 【2】

innovation 改革，创新 【15】

innovation culture 创新文化 【17】

insight 洞察力，见识 【22】

inspiration 灵感 【6】

inspirit 激励 【8】

instinct 本能 【19】

integrative design 整合设计 【21】

instinctive feel 本能的感觉，直觉 【19】

intelligence *n.* 智力，聪明，智慧 【20】

integrate *v.* 结合

vt. 使成整体，使一体化

integrate... into (...) 把……综合为(……) 【23】

integration *n.* 综合

integration of... into(...) ……与(……)的整合 【23】

interaction 交互作用，互动作用 【24】

interpreter　翻译者，口译人员　【16】

intimate　亲密的，隐私的；熟友　【22】

intuition　直觉　【6】

(be)irrelevant (to)　与……不相关　【18】

J

(无)

K

(无)

L

landscape　景观，风景　【22】

laptop　膝上型电脑　【12】

laser　激光　【12】

laud　称赞，赞美

be lauded as...　被赞誉为……　【11】

launch　投放市场；发射　【8】

LCD screen　液晶屏　【10】

leave　*n.*　许可，同意，请假，休假

　　vt.　离开，动身，遗忘，剩下，委托

leave... for (...)　离开……去(……)

　　把……留给(……)

　　vi.　出发，动身，生叶　【20】

library research　图书馆(资料)研究　【14】

life-span design　全寿命设计　【21】

literal meaning　字面含义　【1】

loyalty　忠诚，忠心　【12】

luxury　豪华的，奢侈的，华贵的　【22】

M

make sense 有意义 【1】

manifesto 宣言，声明 【8】

market *n.* 市场，销路，行情

vt. 在市场上交易，使上市

vi. 在市场上买卖 【1】

market need 市场需求 【13】

market needs research 市场需求调查 【14】

market pull 市场吸引 【15】

market research 市场调研 【13】

marketing staff 市场营销人员 【7】

mass 块；群众；大量，大规模的 【22】

massively 严重地，结实地 【12】

matter *n.* 事件，问题，物质，内容，实质，原因，文件

vi. 有关系，要紧 【11】

mean(过去时 meant) (to) 打算，有意要 【5】

media 媒体 【10】

metaphor 隐喻 【5】

metaproject 变形设计，变形规划 【8】

methodology 方法学，方法论 【23】

mission 使命，任务 【23】

modesty 谦逊，虚心 【16】

monopoly profit 垄断利润 【17】

monster 巨人，怪物 【16】

moral 道德；精神上的 【22】

more than ever 益发，愈加 【12】

more than half the battle 成功了一多半 【1】

motivation 动机 【11】

motive *n.* 动机，目的 【11】

adj. 发动的，运动的 【11】

moulding 浇注，浇铸，模铸 【12】

mountain bike 山地自行车 【10】

multicultural 多元文化的 【10】

multimedia 多媒体 【10】

N

no matter how 无论怎样 【11】

non-discrimination 无歧视 【21】

O

opaque 不透明的 【12】

opportunism 机会主义者 【17】

opportunity *n.* 机会，时机 【13】

opportunity specification 机会规范，机会特性 【13】

optical frames 眼镜架 【16】

option 选项，选购件；选择权 【10】

orient *n.* 东方，东方诸国

adj. 东方的，上升的，灿烂的

vt. 使朝东，使适应，确定方向 【18】

orthographic analysis 正交分析(法) 【5】

out of the blue 出乎意外 【6】

overwhelming 压倒性的，无法抵御的 【10】

ownership 所有权，物主身份 【24】

P

paradigm 范例 【10】

parametric analysis 参数分析法 【20】

party animal 社交(动物)一族 【2】

passion *n.* 激情，热情 【12】

passionate 充满热情的 【12】

patent *n.* 专利权，执照，专利品

adj. 特许的，专利的

vt. 取得……的专利权 【17】

patent protection 专利保护 【17】

people with disability disables 残障者 【21】

perception *n.* 知觉，知觉力 【2】

perfect *n.* 完成式

adj. 完美的，理想的，正确的

vt. 使完美，修改，使熟练 【19】

perfection 完整性，尽善尽美 【19】

persona 人，角色 【24】

personal empowerment 个人权利 【21】

pewter 锡镴，白镴 【8】

philosopher 哲学家，哲人 【16】

philosophy *n.* 哲学，哲学体系，达观，冷静 【16】

pioneering strategy 先锋型策略 【17】

pitch in 努力投入工作 【11】

plated *adj.* 镀金的，装甲的 【8】

point *n.* 点，尖端，分数，要点

vt. 弄尖；指向；指出；瞄准，加标

vi. 指，指向，表明 【11】

polymer 聚合物 【12】

position *n.* 位置，职位，形势，阵地

vt. 安置，决定……的位置 【13】

practical *adj.* 实际的；实践的，实用的，应用的，有实际经验的 【5】

practicality 实用性 【5】

pre-attentive 前期注意 【3】

pre-conception 前期概念 【3】

predecessor 前任，前辈 【1】

predispose to 倾向于 【4】

premise　前提，假定　【23】

price elasticity　价格弹性　【20】

price position　价格定位　【13】

prime　*n.*　最初，青春，精华

adj.　主要的，最初的，根本的，一流的，有青春活力的

v.　预先准备好，[口] 让人吃 (喝)足，灌注，填装　【18】

prime function　主要功能　【18】

principles of creativity　创造性原理　【20】

prior knowledge attractiveness　前理解的魅力　【1】

priority　优先次序　【17】

pro-active strategy　前摄型策略　【17】

problem abstraction　问题提取　【20】

problem boundary　问题边界　【6】

problem definition　问题定义　【7】

problem gap　问题间隙　【6】

procedure　*n.*　程序，手续　【5】

product competing strength　产品竞争力　【14】

product development strategy　产品开发战略　【15】

product differentiation　产品的差异化　【20】

product feature　产品特性　【5】

product function　产品功能　【5】

product function analysis　产品功能分析　【18】

product innovation　产品创新　【15】

product marketing　产品营销　【1】

product meanings　产品的含义　【2】

product opportunity　产品机会　【14】

product planning　产品企划　【2】

product semantics　产品语义学　【2】

product survey　产品调研　【14】

product's symbolism　产品符号学　【2】

profit　*n.*　利润，益处，得益

vi. 得益，利用 【17】

project *n.* 计划，方案；事业，企业；工程

v. 设计，投射放映，发射 【8】

prone(to) 倾向 【19】

prototype 原型，样机 【24】

proverb 箴言 【5】

provoker 煽风点火者 【16】

proximity 近似性 【4】

prudence 审慎；节俭 【22】

psychologist 心理学家 【23】

pursue 追赶，追踪，持续 【8】

push *n.* 推，推动；奋发，进取心

vt. 推，推动，推行 【15】

Q

qualitative *adj.* 性质上的，定性的 【14】

qualitative survey 定性调查 【14】

quality 质量，品质，性质 【13】

quality control 质量控制 【13】

quantitative 数量的，定量的 【14】

quantitative survey 定量调查 【14】

questionnaire 问卷，调查表 【14】

R

rank 排队，排序 【7】

rebut 反驳 【24】

reference 提及，涉及，参考，参考书目，证明书(人)，介绍信(人) 【7】

reference concept 参考概念，参照概念 【7】

relative merit 相对价值 【7】

relative *n.* 亲戚，关系词；相关物

adj. 有关系的，相对的 【7】

reliance 信心，依靠，依靠的人或物 【12】

(be)reliant on 依赖于，依靠于 【12】

relationship(with) 关系，关联 【24】

renown 名声，声望 【16】

repeat sales 重复销售 【1】

research *n.* 研究，调查

vi. 研究，调查 【13】

response *n.* 回答，响应，反应 【22】

responsibility 责任(性)，职责 【22】

responsive strategy 响应型策略 【17】

revenue 收入，税收 【11】

rigour 严格，严峻 【8】

risk management 风险管理 【17】

ritual 仪式(的)，(宗教)典礼(的) 【8】

rule *n.* 规则，惯例，章程，标准，控制

vt. 规定，统治，支配，裁决 【3】

rule of symmetry 对称性规则 【4】

S

sales and distribution channels 销售分配渠道 【9】

sales survey 销售调查 【14】

satisfaction 满意，满足，令人满意的事物 【24】

saviour 救世主，救星 【19】

scenario 情节 【24】

sculpture 雕塑 【11】

seamlessly 无缝的，紧密无间的 【23】

selection criteria 选择标准 【7】

sense *n.* 官能，感觉，判断力，见识，意义，理性

vt. 感到，理解，认识 【1】

sensorial 感知的，感觉的 【10】

set-up 组织机构，结构 【12】

shareholder 股东 【23】

shoot(ing) 射击；注射，注塑 【12】

sign *n.* 标记，符号；记号，迹象，征候

v. 签名(于)，署名(于)；签署 【2】

silver-plated 镀银的，包银的 【8】

simplicity *n.* 简单，简易，朴素，直率 【4】

similarity 相似性 【4】

slim 苗条的，纤细的，小巧的 【12】

soar 剧增，高飞 【11】

sociologist 社会学家 【23】

solution *n.* 解答，溶解，溶液；解决方案 【6】

sound *n.* 声音，噪声，吵闹，海峡

adj. 可靠的，合理的，有效彻底的 【9】

sounding 收集意见，调查(结果) 【9】

sound out 试探，调查 【9】

speciality 特性，特质，专业，特殊性 【14】

specification 详述 [常 pl.]，规格，说明书，规范 【13】

spoil 扰乱，搞糟 【11】

staff *n.* 棒，杖杆，支柱，全体职员

vt. 供给人员；充当职员 【7】

stairway to creativity 创造性阶梯 【20】

static market 静态市场 【17】

statutory 法律法规的 【9】

sticky dot 带背胶的圆点贴纸 【7】

stigma 污名，耻辱 【10】

stimulus 刺激物，促进因素；刺激 【8】

strategic planning 战略性规划 【17】

strategics *n.* 兵法，战略学 【17】

strategy *n.* 策略，战略 【15】

strategy for product development　产品开发策略　【17】

straying　迷路的，离群的　【16】

strike　*n.*　罢工，打击，殴打

　　vt.　(过去时struck)打，拍板，冲击，罢工；打动　【11】

structive　结构的，建筑的　【22】

studio　(设计)工作室　【10】

style　*n.*　风格，时尚，文体，风度，类型

　　vt.　称呼，设计，使合潮流　【2】

styling principle　造型原则　【6】

subsidiary functions　辅助功能，补充功能　【18】

suffer a frustration　遭受挫折　【1】

suitability(for)　(与……的)适应(适宜)性　【9】

suitability　合适，适当，相配，适宜性　【9】

survey　*n.*　测量，调查，

　　vt.　调查(收入，民意等)，测量　【14】

survival　生存，幸存；幸存者　【22】

sustainable　可支持的，可维持的，可持续(发展)的　【22】

switch　*n.*　开关，电闸，转换

　　vt.　转换，转变　【11】

symbiosis　共生(现象)，互利合作关系　【22】

symbolic attractiveness　符号魅力　【2】

symmetry　对称，匀称　【3】

synonymous (with)　(与……)同义的　【11】

T

tangible　切实的　【15】

tangible result　有形结果　【15】

target　目标，对象，靶子　【14】

target cost　目标成本　【13】

target market　目标市场　【14】

team-working 团队工作 【19】

technology push 技术推动 【15】

technology auditing 技术审核 【13】

tend to 倾向于，趋于 【4】

theme board 主题看板 【2】

there’s no point 毫无意义 【11】

think of... as... 把……看做是……，认为……是……

(be)thought of as 看做…… 【24】

time and again 屡次，反复不断地 【11】

time to market 上市时间 【17】

to some extent 在某种程度上 【17】

to the full 充分 【20】

traditional strategy 传统型策略 【17】

transgenerational design 跨代设计 【21】

translator 翻译者，解释者 【16】

trustee 托管人，保管人 【16】

twin-shot 双塑注射，双料注塑 【12】

typology 形体学，象征学；血型学 【8】

U

undulating 波浪形的 【16】

universal 普遍的，全体的，通用的，宇宙的，世界的 【21】

universal design 通用设计 【21】

user-centeredness 用户中心论 【21】

usefulness 有用，有效性 【24】

utopian 乌托邦的，理想化的 【8】

V

vacate 腾出(空间，职位)，退位 【8】

validate 使有效，使生效，确认，证实 【24】

verification　查证确认　　【20】

veritable　真正的　　【8】

viability　生存能力，生存性，发育能力　　【15】

view　*n.*　景色，风景；观点，见解；观察，观看；意见，认为

　　vt.　观察，观看　　【10】

village community　乡村社区　　【2】

virtuoso　大师，名家，学者　　【19】

vision　*n.*　视觉 先见之明，视力，眼力 想象力，先见之明，幻想，景象

　　vt.　梦见，想象，显示　　【1】

visual appearance　视觉外观　　【1】

visual harmony　视觉协调　　【4】

visual image　视觉形象　　【3】

visual information　视觉信息　　【3】

visual perception　视觉感知　　【3】

visual process　视觉过程　　【3】

visual signal　视觉信号　　【3】

visual simplicity　视觉简单性　　【4】

visual styling　视觉形态　　【3】

visual theme　视觉主题　　【4】

W

ware　陶器，器皿　　【11】

wearable　耐久的　　【10】

weld(ing)　焊接；焊缝　　【12】

what if...?　如果……那将如何呢？　　【19】

wireframing　线框图(法)　　【24】

with regard to　关于　　【17】

wondrous creature　非凡的人物，精英　　【19】

注：在词汇及其中文意义后的方括号【】中的数字表示词汇在本书内首次出现的课数。

后　记

虽然也参与过工业设计专业英语教材的编写，但总觉得现今的教材还离不开普通英语教科书的套路，往往很少考虑专业本身系统及语言表达上应有的特点。有了这种潜意识，当遇到外校的几位老师谈及编写中英双语教材时就把过去的甜酸苦辣忘了个一干二净，痛快地答应了。熬了许久，现在脱稿了，要付印了。当然其间还有许多人的鼎立相助，没有这些鼎力相助恐怕要熬也是熬不过去的。

首先要感谢的是主要参加编写的华东师范大学设计学院的毛溪女士。毛溪的本科和硕士研究生都是在江南大学设计学院就读的，毕业后到华东师范大学设计学院任教。从考虑主旨到编排版面等方面，她无事不为。尤其是当我这个书呆子抱着那些“高深莫测”的专业材料当宝贝不放时，她提醒我要懂得取舍：凡是不适合既定读者群的内容再“好”也必须舍弃。于是才有了今天可以呈献给大家的这本书的现在这个样子。

还要感谢我夫人毛荫秋。作为有工科背景(她原是学热工量测与自动化专业的)的英语翻译(也是由于家庭原因要回老家无锡、被迫改行当翻译)，她对工业设计原本并不熟悉。但经过这十来年在无锡市举办的各种工业设计节(周)活动以及参与江南大学设计学院的许多国际学术交流活动(尤其是由我负责的江南大学蒋震基金工业设计培训中心举办的许多国际学术交流、讲学和培训活动)的锻炼，专业知识已不外行，在业内也有了点小“名气”。她的英语比我强，这次在编写这本书时着实地助了我一臂之力。

当然，还必须提及江南大学设计学院的许多新老领导、老师。他们不仅在原则上给我指点和支持，还在许多方面给予了具体的指导和帮助。在此我要对他们表达由衷的感谢！

此外，在本书撰稿的一段时间里华东师范大学设计学院的不少学生在毛溪老师的指导下也在多方面参与了本书的编写工作。而我甚至都没有与他们谋过面。对他们的辛勤劳动，在此深表感谢！

热心人无处不在。本书的一般读者也许都不会想到，这本书里还有美国缅因州一位美国朋友的贡献。当他得知我在某个部分的编写有困难时，立即给我发来了相关资料。我只能在心里感谢他。

最后，还要感谢为本书的推介作出积极评价的江南大学张福昌教授、浙江大学许喜华教授、南京艺术学院何晓佑教授及上海视觉艺术学院张同教授，感谢他们的鼓励和帮助。

由于我和其他编写者的能力、水平和见识有限，本书中难免会有不少错误。恳请各位专家、同行、老师、学生和所有读者不吝赐教，直率地给我们指正。我们将在重印、修订或再版时尽量予以改正。

江建民

2009年7月于无锡